39,95
68I

# ENGINEERING ASPECTS OF WATER LAW

# ENGINEERING ASPECTS OF WATER LAW

LEONARD RICE
President
Leonard Rice Consulting Water Engineers, Inc.

MICHAEL D. WHITE
Partner
White & Jankowski

KF
5569
.R53
1987
West

A Wiley-Interscience Publication
**JOHN WILEY & SONS**
New York / Chichester / Brisbane / Toronto / Singapore

Arizona State Univ.
West Campus Library

Copyright © 1987 by John Wiley & Sons, Inc.

All rights reserved. Published simultaneously in Canada.

Reproduction or translation of any part of this work beyond that permitted by Section 107 or 108 of the 1976 United States Copyright Act without the permission of the copyright owner is unlawful. Requests for permission or further information should be addressed to the Permission Department, John Wiley & Sons, Inc.

***Library of Congress Cataloging in Publication Data:***

Rice, Leonard.
Engineering aspects of water law.

"A Wiley-Interscience publication."
Bibliography: p.
Includes index.
1. Water resources development—Law and legislation—United States. 2. Water-rights—United States.
I. White, Michael D. II. Title.
KF5569.R53 1987 346.7304'69115 86-22415
ISBN 0-471-84362-8 347.306469115

Printed in the United States of America

10 9 8 7 6 5 4 3 2 1

# PREFACE

This book is an introductory guide for students, laymen, and beginning professionals in the water resources area. It is based on extensive teaching experience with law students, lawyers, and undergraduate and graduate engineers. It was prepared in large part from our teaching notes and primarily in response to the need for an introductory text of some breadth—in other words, a book that covers the subject matter but does not belabor the details.

Those using this book must be prepared to supplement it with information which is specific to their locale. The acute reader will undoubtedly notice the frequent use of words such as "often," "some," "many," and so on. It was not our intent to quibble about the facts of life. To the contrary, we are well aware that there are few hard and fast rules in the water business. Even those that appear to be cast in concrete in one state will be quite different in another.

Throughout the book we try to avoid being drawn into tangents. As any water professional will know, few water lawyers are able to escape involving themselves in virtually every area of the law. Similarly, water engineers find themselves working in obscure areas of engineering and the physical sciences. Nevertheless, we made a concerted effort to not limit our discussion to traditional water law subjects and to keep that discussion on a very general plane.

It is also probably true that, since this was intended primarily as an engineering text, the engineering aspects receive far more sophisticated treatment than do the legal considerations. For those who are uncomfortable with such cavalier treatment of somber legal matters, we would suggest that the book be supplemented with more detailed material on local law or a purely legal work such as Getches, *Water Law in a Nutshell.*

Finally, insofar as possible in a general work, we try to emphasize the practical as opposed to the theoretical. This too may rub our academic colleagues the wrong way. Yet few other areas of the law or engineering are so dominated by practicalities. The water resources business, when viewed from a practical and historical perspective, is logical and enjoys reasonable symmetry. The greatest difficulty faced by newcomers to the field is that they can expect too much since certainty and consistency are never anticipated in the combination of the law of the physical resources and of the perversities of human nature.

Much of the history and future of the western United States are directly tied to water resources. Consequently, the water business cannot be viewed solely from a legal or an engineering standpoint. To get a taste of the entire substantive area, other disciplines such as geography, economics, political science, sociology, soil science, and agriculture must be included in the inquiry.

What we have produced is nothing more than a starting point. Good luck in the remainder of your journey.

LEONARD RICE
MICHAEL D. WHITE

*Denver, Colorado*
*January 1987*

# CONTENTS

# ENGINEERING ASPECTS OF WATER LAW

# 1

# INTRODUCTION

A river is more than an amenity, it is a treasure. It offers a necessity of life that must be rationed among those who have power over it.

OLIVER WENDELL HOLMES

Once the province of only lawyers and the courts, water law* has dramatically changed during this century through the influence of engineering and the physical sciences. Future changes are inevitable as concepts from the social sciences assume increasing importance. By necessity, therefore, the water community is becoming increasingly interdisciplinary.

To thoroughly understand and effectively participate in the water business, one must be at least conversant in a multitude of disciplines. Since that does not happen overnight and usually requires years of practical experience, let's start with the fundamentals: the law, the resource, and the application of engineering principles.

## IMPORTANCE OF WATER RIGHTS

A "water right" is simply the legal right to use water. In the eastern United States, the right has historically existed only for land that is adjacent, or riparian, to a stream. The riparian doctrine worked satisfactorily in climates where water was plentiful and uses relatively small and nonconsumptive.

*Water law is the creation, allocation, and administration of water rights.

Where water was scarce and unevenly distributed, as in the western United States, the riparian doctrine did not work well. Limiting the use of water only to lands next to streams proved quite impractical. In the rough and tumble early West, water rights were created merely by taking (or appropriating) the water and using it, regardless of the location or ownership of its place of use. The only limitation on the use was that it be "beneficial," usually based on utilitarian concepts. With refinements, the appropriation doctrine continues to be the basis for allocation of water in most of the West.

Appropriative water right concepts are important because they combine to form a system for allocating an essential resource in a manner that can be administered for the public good. Through practical application of the principles that relative priority of appropriation establishes the basis for allocation, that beneficial use is the measure of the right, and that a water right is a property right which (absent of injury to other water rights) can be transferred and changed in place and type of use, the system tends to prevent waste and encourages economic development through operation of the free market. In some states water rights are also used to establish minimum streamflows and lake levels which serve not only to maintain water quality but to preserve aesthetic and recreational values.

## THE HYDROLOGIC CYCLE

Water is a renewable resource and differs from other resources such as coal and oil in that it is not destroyed by use, although it may be altered in quantity, quality, and physical state. Essentially, the volume of water in the earth's atmosphere is fixed, but its distribution is constantly changing in place and physical state by movement through the hydrologic cycle.

The hydrologic cycle, depicted in Figure 1, describes the process by which water changes from a liquid to a vapor through evaporation from water surfaces and transpiration by plants, is then condensed into clouds and fog, and eventually returns to the earth as precipitation in the form of rain, snow, sleet, and hail. Water appears on the earth's surface in many forms, from glacial ice to flowing rivers and tranquil lakes. Water also occurs below the earth's surface at various depths, from shallow sand and gravel deposits in river valleys, to deep underground formations. These are both known as aquifers. Subsurface water moving through the earth recharges rivers and lakes, as part of the hydrologic cycle.

## THE WATER RESOURCE

### Problems with Distribution

Water resource management often involves dealing with problems caused by the unequal geographic and temporal distribution of water. Although the total volume of water contained within the earth's atmosphere and on and below the

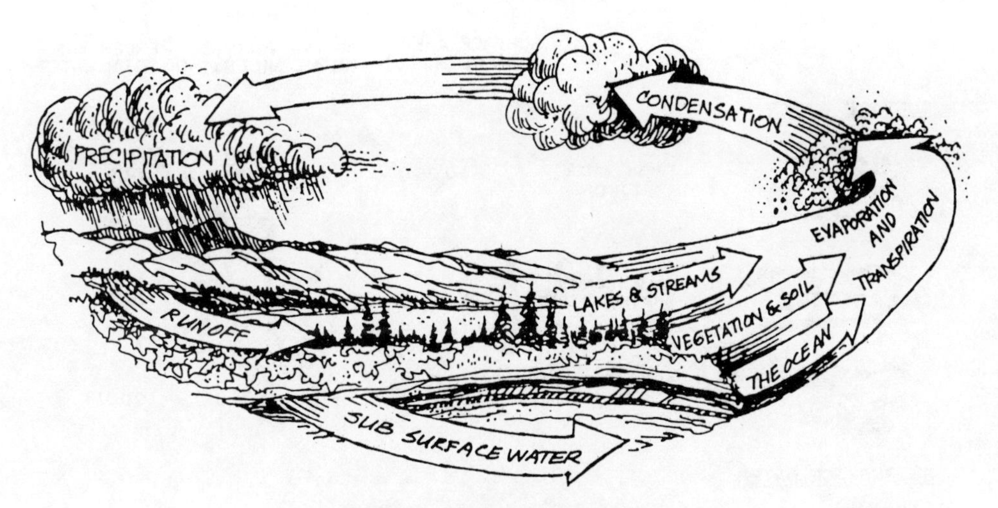


**FIGURE 1.** The Hydrologic cycle.

earth's surface is large, as shown in Figure 2, only a relatively small portion is available for development and use, and often not at the time or place where needed. In Colorado, for instance, the major portion of the state's water supply originates in the mountains and runs off to the western side of the Continental Divide. However, the largest population centers and most extensive agricultural areas are on the eastern slope of the Continental Divide. This has historically caused major conflicts between the sparsely populated and agrarian west slope and the urbanized east slope, as water has been diverted from west to east. A similar situation exists in California where the water supply largely originates in the north but is diverted and conveyed to the urban areas of the south.

In total, the United States has abundant water resources with average annual precipitation of 30 inches for the contiguous United States, average natural runoff of 3.7 million acre-feet per day, and large reserves of underground water. The situation is less favorable in terms of distribution and timing of the water resource. Rainfall varies from abundant in the Pacific Northwest to extreme scarcity in the Southwest. In any one region, streamflow can vary widely from season to season and from year to year. Figure 3 shows the variation in runoff from year to year and from season to season for a typical western stream.

## Precipitation

Approximately 10 percent of the atmospheric moisture over the United States reaches the ground as rain or snow. Evaporation from oceans provides over 80 percent of this precipitation, and less than 20 percent is supplied from inland evaporation and transpiration from plants.

East of the Rocky Mountains, the major moisture sources are the Gulf of Mexico and the Atlantic Ocean, while west of the Rockies, the main source is the Pacific Ocean. Significant amounts of water are precipitated in local areas on the wind-

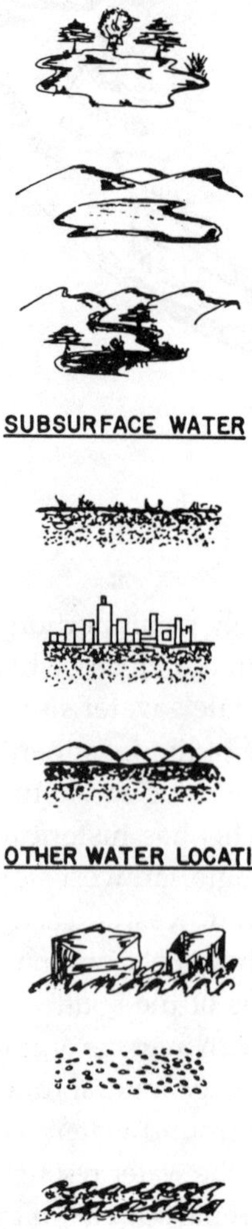

| | LOCATION | SURFACE AREA (SQUARE MILES) | WATER VOLUME (CUBIC MILES)* | PERCENTAGE OF TOTAL WATER |
|---|---|---|---|---|
| **SURFACE WATER** | | | | |
| | FRESH-WATER LAKES | 330,000 | 30,000 | .009 |
| | SALINE LAKES AND INLAND SEAS | 270,000 | 25,000 | .008 |
| | AVERAGE IN STREAM CHANNELS | — | 300 | .0001 |
| **SUBSURFACE WATER** | | | | |
| | VADOSE WATER (INCLUDES SOIL MOISTURE) | 50,000,000 | 16,000 | .005 |
| | GROUND WATER WITHIN DEPTH OF HALF A MILE. | | 1,000,000 | .31 |
| | GROUND WATER-DEEP LYING | | 1,000,000 | .31 |
| **OTHER WATER LOCATIONS** | | | | |
| | ICECAPS AND GLACIERS | 6,900,000 | 7,000,000 | 2.15 |
| | ATMOSPHERE (AT SEA LEVEL) | 197,000,000 | 3,100 | .001 |
| | WORLD OCEAN | 139,500,000 | 317,000,000 | 97.2 |
| | TOTALS (ROUNDED) | | 326,000,000 | 100 |

* ONE CUBIC MILE OF WATER EQUALS 1.1 TRILLION GALLONS

SOURCE: US DEPARTMENT OF THE INTERIOR/ GEOLOGICAL SURVEY

**FIGURE 2.** Distribution of world's estimated water supply.

ward side of the high mountains as the atmospheric moisture is forced to rise as it moves inland.

The annual precipitation within the 48 contiguous United States ranges from less than 4 inches in the Great Basin to more than 200 inches in the northwest coastal areas, with an average of approximately 30 inches. Approximately 26

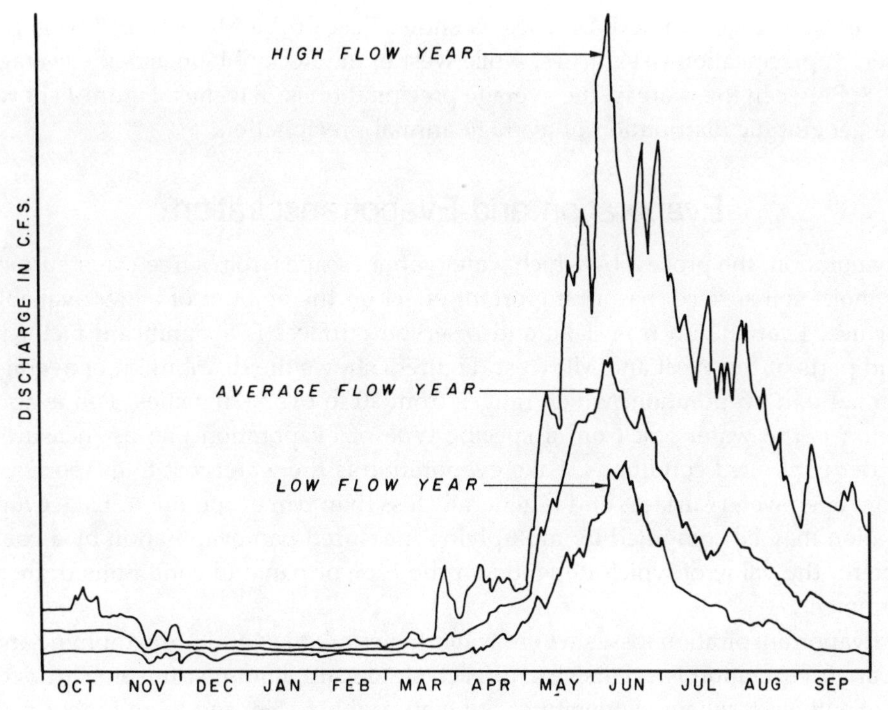


**FIGURE 3.** Variations in annual and seasonal runoff.

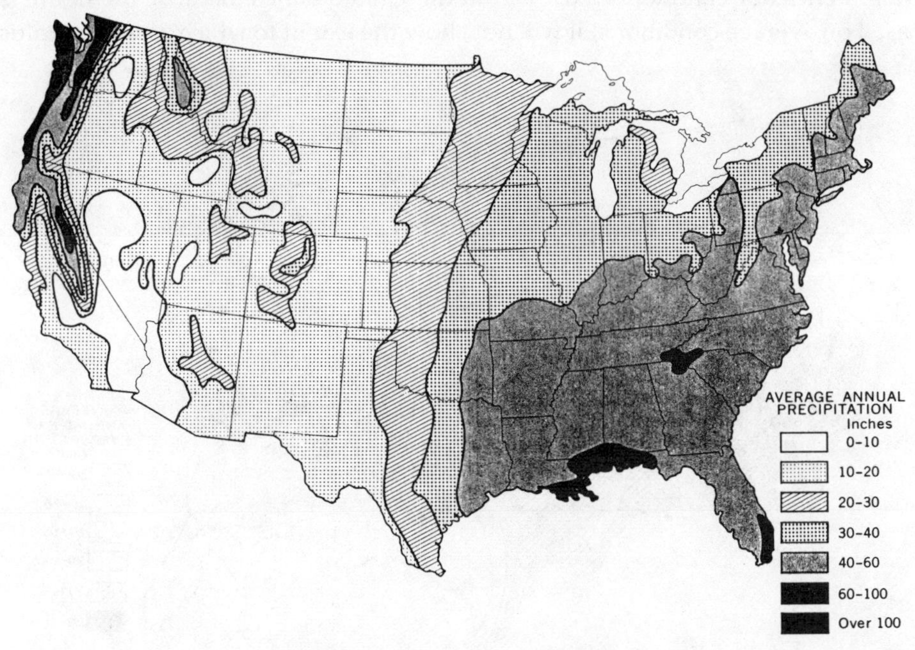


**FIGURE 4.** Average annual precipitation.

inches occurs as rain and 4 inches as snow. East of the Mississippi River, the average precipitation is 44 inches, while west of the Rocky Mountains the average is 18. Between these areas, the average precipitation is 28 inches. Figure 4 shows the geographic distribution of average annual precipitation.

## Evaporation and Evapotranspiration

Evaporation, the process by which water vapor escapes from a free water surface or moist soil surface, has an important effect on the amount of water available for use. Evaporation from lake and reservoir surfaces is a significant factor in arid parts of the West and Midwest. Figure 5 shows the distribution of average annual pan evaporation, which ranges from 20 to over 120 inches. Pan evaporation is the water lost from a specific type of evaporation pan as measured under controlled conditions. Lake evaporation is the water lost to evaporation from open water surfaces and is generally less than pan evaporation. Lake evaporation may be computed by multiplying measured pan evaporation by a coefficient, the value of which depends on the type of pan and conditions of measurement.

Evapotranspiration losses where water tables are high and phreatophytic and riparian vegetation is extensive can reach significant amounts. In some areas of the Southwest, where temperatures are high, soil is fertile, and there is abundant subsurface moisture, losses of up to 80 inches annually have been recorded.

Figure 6 was prepared by subtracting values of potential evapotranspiration from average precipitation, and shows areas in which natural water surplus or water deficiency commonly exist within the United States. Because the figure is based on average conditions, it will not show the extent to which annual surplus

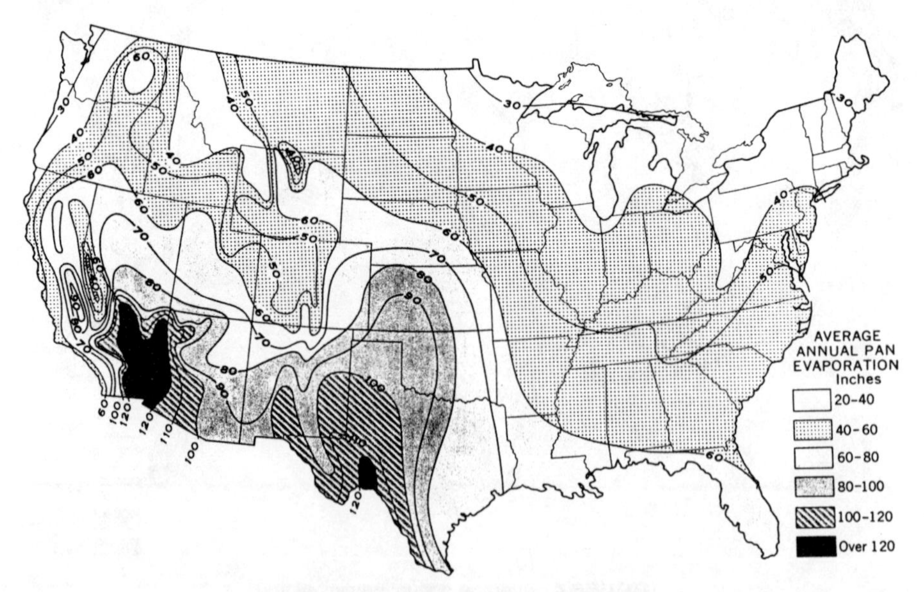


**FIGURE 5.** Average annual pan evaporation.

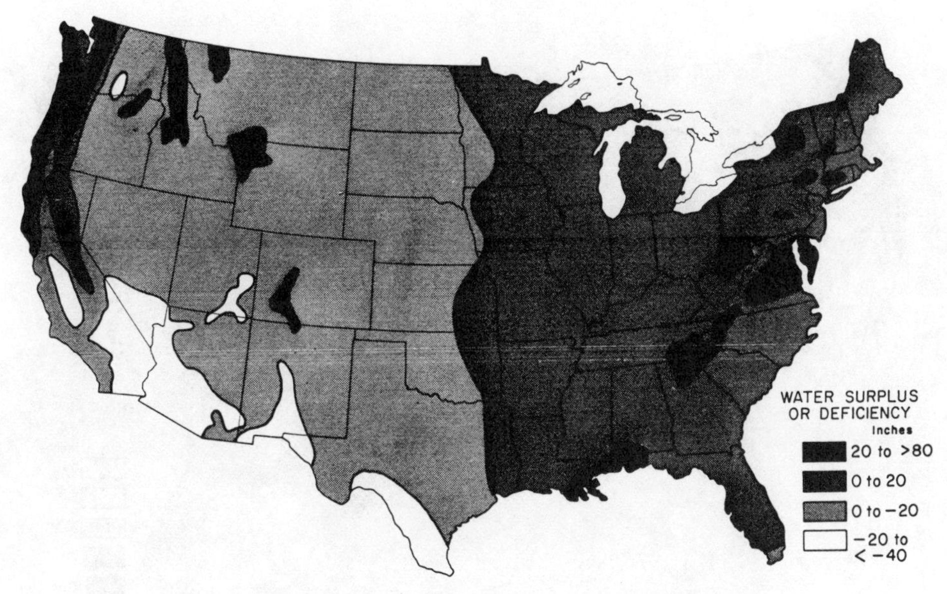


**FIGURE 6.** Areas of natural water surplus and deficiency.

or seasonal short-term deficiencies may occur within any region. The map shows that a natural water surplus occurs in the eastern half of the United States with a general natural water deficiency in the western half, excluding the northwest and west coast.

## Surface Water

Surface streamflow is highly variable and provides a major portion of the usable freshwater resource. Rivers and streams supply about 67 percent of the water used, and it is not unusual within a normal year for the ratio of maximum flow to minimum flow to be as much as 500 to 1. As shown by Figure 3, year to year variations in flow can be substantial, as can seasonal variations. The amount of water available from streams varies greatly in different parts of the country and is directly related to the precipitation, evaporation potential, and topographic characteristics of the basin. Figure 7 shows the average annual runoff in the United States based on data for the period 1931–1960. In much of the western United States, the average annual runoff is less than 1 inch, whereas it may exceed 10 to 20 inches in the eastern United States.

The variability of annual natural runoff is illustrated in Tables 1 and 2. Table 1 shows the annual runoff available 50 percent, 90 percent, and 95 percent of the years and is based on statistical distributions of annual runoff at 700 gaging stations in the continental United States. The table shows the difference in variability of natural runoff between the various regions. For example, in the north Atlantic region the annual runoff tends to vary least from year to year and the annual runoff exceeded 95 percent of the years is 69 percent of the average. In

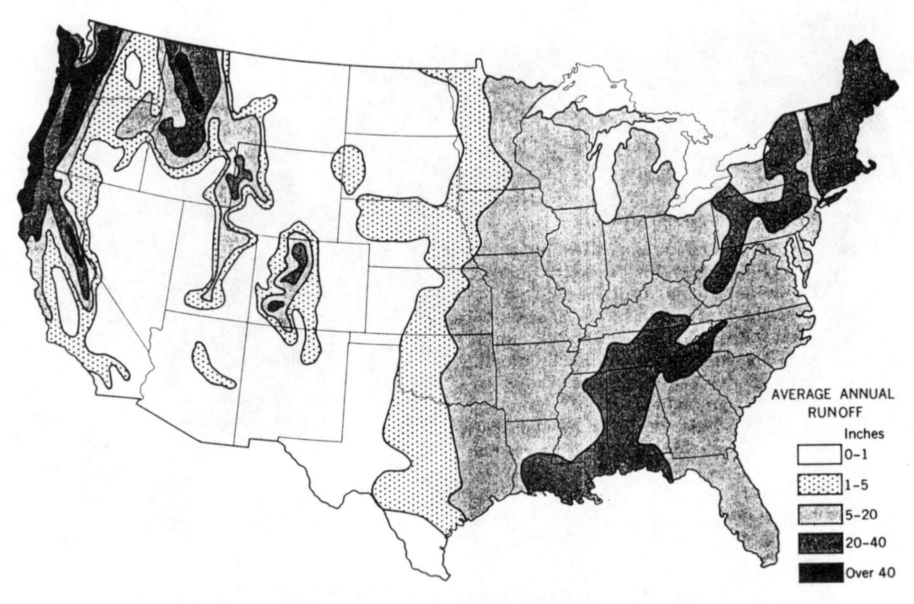


**FIGURE 7.** Average annual runoff.

the lower Colorado region where the annual runoff varies widely from year to year, the annual runoff exceeded 95 percent of the years is only 25 percent of the average.

Seasonal runoff is related to the source of water and the location, with the general pattern being high flows during spring and low flows during late summer and early winter.

Precipitation and surface runoff are also subject to long term fluctuations that can have significant impacts on water resource policies and management. An example of this is the Colorado River Basin, which is subject to a combination of interstate compacts, court decrees, international treaties, and Federal statutes that collectively form the Law of the River. During the first third of the twentieth century, the annual runoff of the Colorado River at Lee Ferry, Arizona, averaged nearly 18 million acre-feet as shown by Figure 8. Since that time the annual runoff has exhibited a decreasing trend and by 1980 averaged less than 14 million acre-feet. Figure 8 also relates the significant events of the Law of the River to the annual virgin flow at Lee Ferry. Of particular interest is the negotiation of the Colorado River Compact in 1922 at a time when the average runoff was near its maximum for this century. This led to a compact allocation of more water than has been in the river since that time and played a significant role in subsequent disputes over the development and use of Colorado River water.

## Groundwater

It is estimated that groundwater supplies approximately 25 percent of the nation's water use. Groundwater also provides the base flow for streams and, in some regions, provides a continuity of discharge that would not otherwise exist.

Table 1
ANNUAL NATURAL RUNOFF
(billions of gallons per day)

| Region | Mean | 50%[a] | 90%[a] | 95%[a] |
|---|---|---|---|---|
| North Atlantic[b] | 163.0 | 163.0 | 123.0 | 112.0 |
| South Atlantic-Gulf | 197.0 | 188.0 | 131.0 | 116.0 |
| Great Lakes[b,c] | 63.2 | 61.4 | 46.3 | 42.4 |
| Ohio[d] | 125.0 | 125.0 | 80.0 | 67.5 |
| Tennessee | 41.5 | 41.5 | 28.2 | 24.4 |
| Upper Mississippi[d] | 64.6 | 64.6 | 36.4 | 28.5 |
| Lower Mississippi[d] | 48.4 | 48.4 | 29.7 | 24.6 |
| Souris-Red-Rainy[b] | 6.17 | 5.95 | 2.60 | 1.91 |
| Missouri[b] | 54.1 | 53.7 | 29.9 | 23.9 |
| Arkansas-White-Red | 95.8 | 93.4 | 44.3 | 33.4 |
| Texas-Gulf | 39.1 | 37.5 | 15.8 | 11.4 |
| Rio Grande[e] | 4.9 | 4.9 | 2.6 | 2.1 |
| Upper Colorado[f] | 13.45 | 13.45 | 8.82 | 7.50 |
| Lower Colorado[d,e] | 3.19 | 2.51 | 1.07 | 0.85 |
| Great Basin[d] | 5.89 | 5.82 | 3.12 | 2.46 |
| Columbia-North Pacific[b] | 210.0 | 210.0 | 154.0 | 138.0 |
| California[e] | 65.1 | 64.1 | 33.8 | 25.6 |
| Conterminous United States[g] | 1,201.0 | | | |
| Alaska[b] | 580 | 8 | 8 | 8 |
| Hawaii | 13.3 | 8 | 8 | 8 |
| United States[g] | 1,794.0 | | | |

[a] Flow exceeded in indicated percent of years.
[b] Does not include runoff from Canada.
[c] Does not include net precipitation on the lakes.
[d] Does not include runoff from upstream regions.
[e] Does not include runoff from Mexico.
[f] Virgin flow. Mean annual natural runoff estimated to be 13.7 billions of gallons per day.
[g] Rounded.
[h] Not available.

Groundwater represents a significant resource in the United States and is estimated to have a volume greater than that of all surface water and more than the total capacity of the nation's lakes and reservoirs. The distribution of groundwater in the United States is shown in Figure 9. The largest reserves of groundwater occur in the Atlantic and Gulf coastal plains, with significant areas of potential groundwater development also located in the alluvial basins in the far West. These consist of alluvium filled valleys surrounded by mountains and are recharged by runoff. Another area of important potential groundwater development exists in the northern part of the country, encompassing the area of glacial deposits which extend from the Rocky Mountains almost to the Atlantic coast, and southward as far as the Missouri and Ohio rivers. This area consists of rock and soil debris left by glaciers and contains beds of water-sorted permeable sand and gravels that constitute an important source of water.

The high plains area is a large remnant of a vast alluvial apron containing

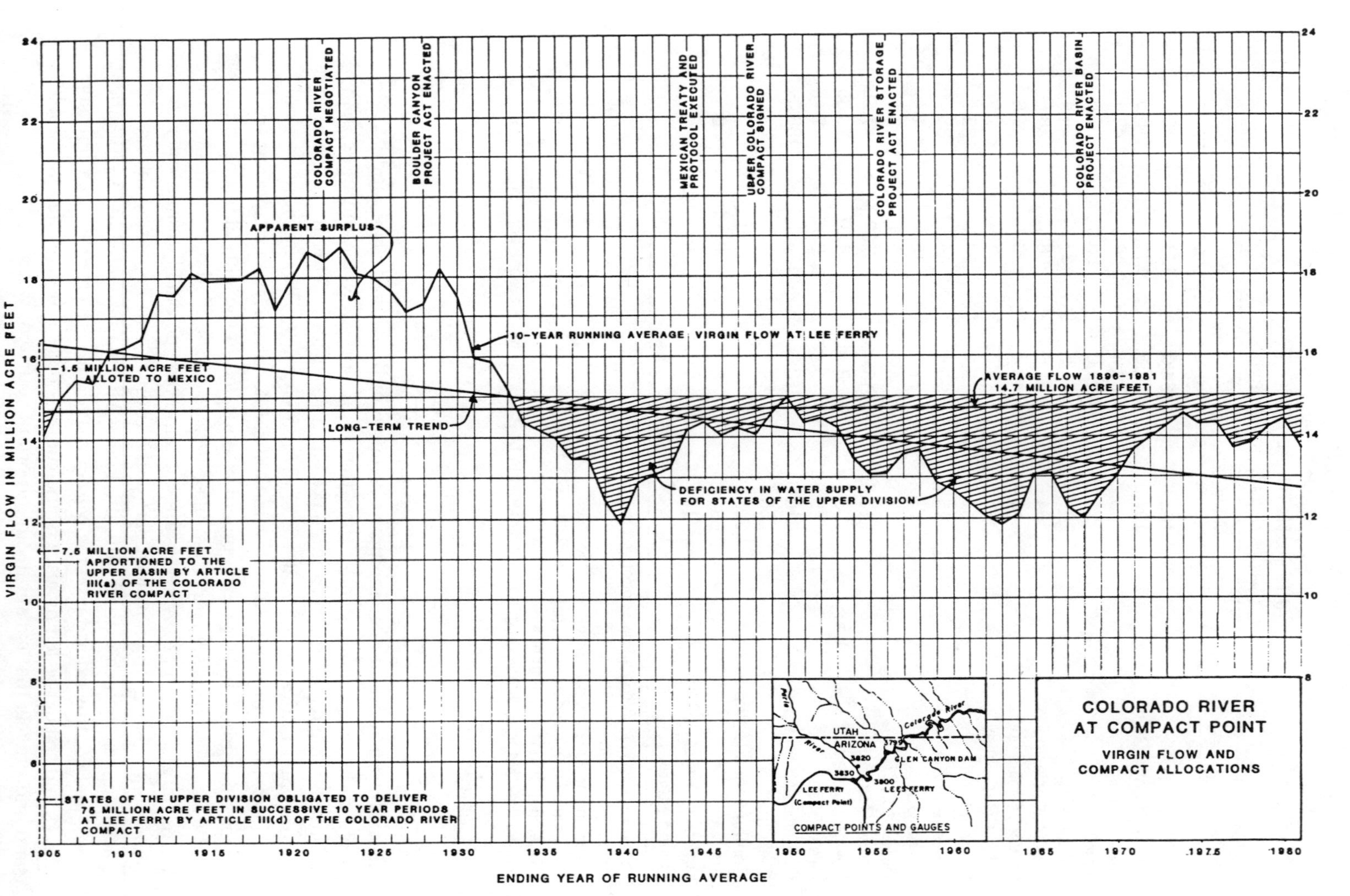


**FIGURE 8.** Colorado River at compact point.

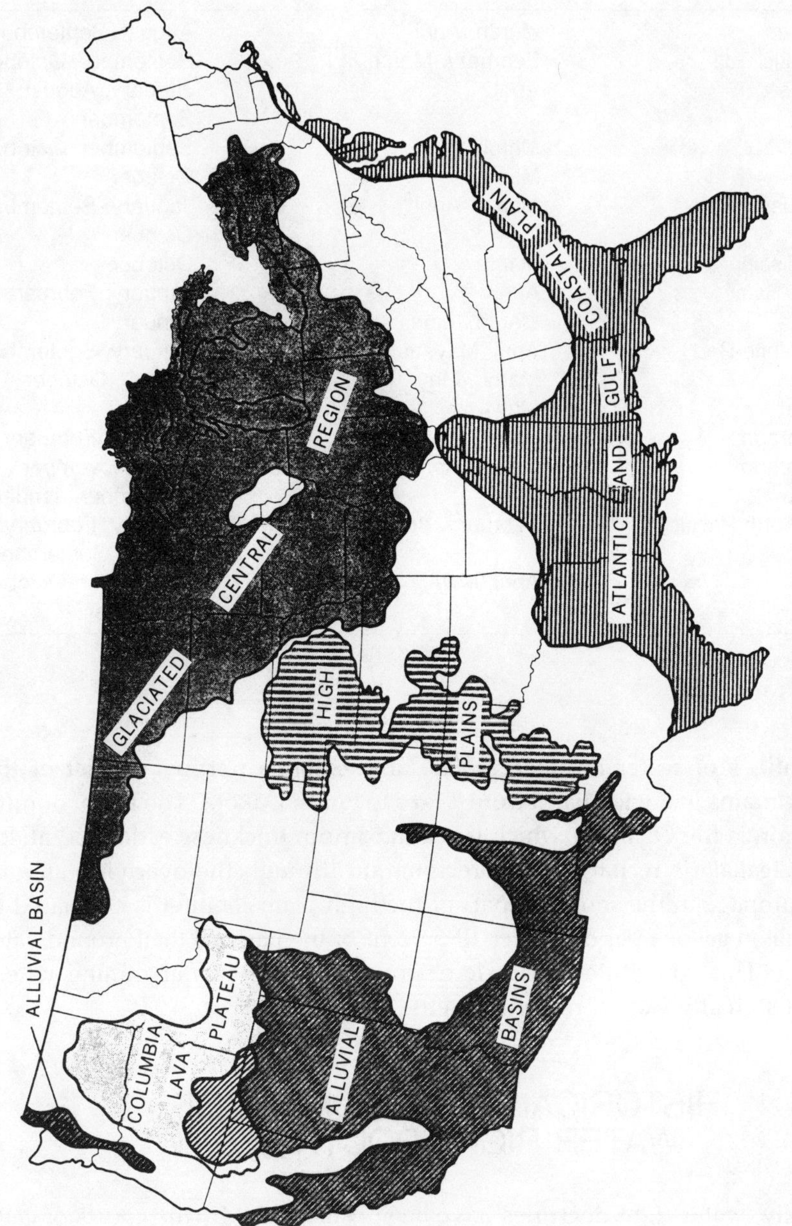


**FIGURE 9.** Areas of groundwater supply.

Table 2
SEASONAL VARIATION OF NATURAL RUNOFF

| Region | Months of High Flow | Months of Low Flow |
|---|---|---|
| North Atlantic | March, April | August, September |
| South Atlantic-Gulf | February, March | September, October |
| Great Lakes | April | January, August, September |
| Ohio | March | September, October |
| Tennessee | March | October |
| Upper Mississippi | March, April | January, September, October |
| Lower Mississippi | March | October |
| Souris-Red-Rainy | April | January, February |
| Missouri | March, June | January |
| Arkansas-White-Red | April, May, June | January, September |
| Texas-Gulf | March, May | August, October |
| Rio Grande | May | June |
| Upper Colorado | June | January, February |
| Lower Colorado | March, April | June, November |
| Great Basin | June | September, January |
| Columbia-North Pacific | February, April, May | January, February, August, September |
| California | April, May | September, October, December |

large quantities of water in storage. The area extends northward east of the Rocky Mountains in a wide band from Texas to South Dakota. The major aquifer of this region is the Ogallala, which has a maximum thickness estimated at 400 feet. The Ogallala is recharged by precipitation through the overlying alluvial aquifer. Pumpage in the southern part of the high plains aquifer is estimated to exceed 7 billion gallons per day, over 10 percent of the national total groundwater withdrawal. This constitutes a classic example of groundwater mining where withdrawals greatly exceed replenishment.

## HISTORICAL DEVELOPMENT OF WATER RIGHT DOCTRINES

Traditionally, water right doctrines have been established by the courts of each *state*—each in different ways and in response to the unique hydrologic conditions prevalent in that state. Although *federal* doctrines have now assumed immense significance, traditional state doctrines still form the backbone of water law today. Nevertheless, one concept of federal law (which establishes each state's share of water) must be addressed before reviewing the various state-created doctrines.

## Equitable Apportionment—The State's Share

In a nation such as ours, where many streams cross state boundaries, there have been a number of unavoidable controversies over how much water from an interstate stream may be allocated to its citizens by any particular state. Those disputes have resulted in the equitable apportionment doctrine, under which, either by interstate compact or by decree of the United States Supreme Court, each state is assigned an equitable share of interstate waters. Once assigned, that share creates a maximum limit on the total amount of water that may be internally allocated by the state. Each state, however, has established its own legal doctrine to allocate its intrastate water, as well as its share of interstate water.

## State Water Right Doctrines

There are at least two major variables affecting state water right doctrines: whether the water involved is found above or below the earth's surface, and whether the state involved is water-short. In either event, the conceptual nature of a water right is the same. It is the right to *use* water, under certain conditions, and it is an interest in real property, for which ownership is treated similarly to ownership for land.

Traditionally, there have been three legal classifications of water: diffuse surface water; water of natural surface streams; and underground water. Since water rights are acquired only for the last two categories and since we tend to be a bit sloppy in our language, there is ample opportunity for confusion. Diffuse surface water is a nuisance. The law's only concern is how people rid themselves of it. Unfortunately, however, common usage is for surface water to mean stream water, at least in the water right context.

### *Streamwater Versus Groundwater*

In the western United States there are several different water right doctrines governing the use of water which may be found in natural streams, or water which may be found beneath the surface of the earth. For our purposes, those can be categorized as streamwater and groundwater.

The historical development of the law regarding these two categories of water has been remarkably different. While many of the physical characteristics and attributes of water found in surface streams were readily understood by virtually everyone, exactly the opposite was true for groundwater. Consequently, disputes over groundwater were rare while disputes over the use of streamwater occurred with surprising frequency. Similarly, the courts were willing to establish legal doctrines for streamwater but were indecisive where the mysteries of groundwater were involved. As a result, by the beginning of the twentieth century, most western states maintained well developed legal doctrines concerning the waters of surface streams. With respect to groundwater, however, those same courts usually left water users to their own devices.

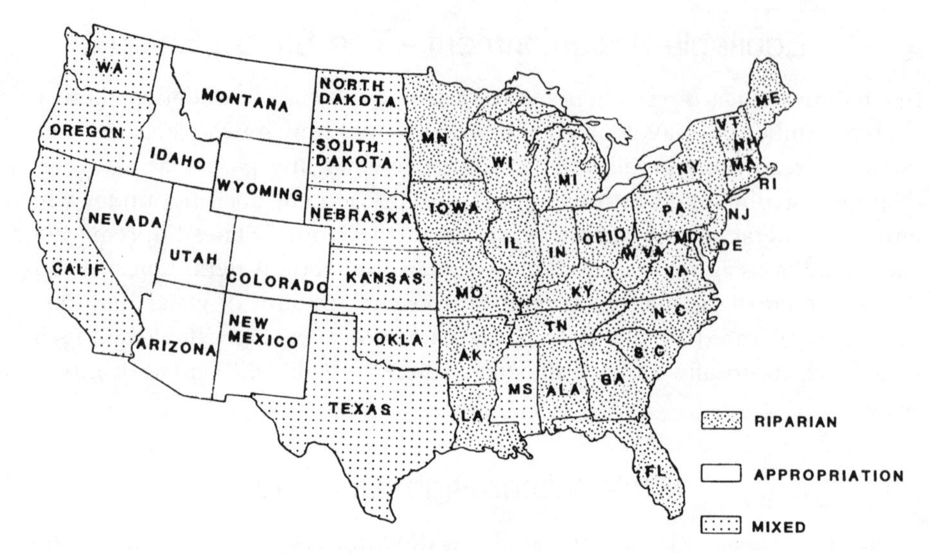


**FIGURE 10.** Predominate streamwater doctrines.

## *Streamwater*

In states with substantial water supplies, courts generally adopted the *riparian* doctrine, allowing water to be used only by owners of land located along a stream under a fairly loose set of legal principles. In many of those states, however, statutes have significantly modified the law.

In dry states, where water was at a premium, courts universally adopted the *appropriation* doctrine, allowing streamwater to be used by anyone, but under a very strict system of priorities. One note of caution is necessary. In some states containing highly divergent climatic regions, both doctrines may have been adopted. Figure 10 shows the distribution of the predominant stream water legal doctrines by state.

*The Riparian Doctrine.* The riparian doctrine had its modern beginnings several centuries ago in England, as a response to early water squabbles between mill owners. Those mills, located on surface streams and rivers, were dependent on the flow of water to rotate water wheels which, in turn, ran machinery to grind grain. Because most rivers were neither particularly steep nor fast-flowing, naturally occurring mill sites were at a premium. The earliest mills were placed at those rare locations where the natural river flows were swift enough to turn a water wheel.

After those particular sites were all occupied, subsequent mills were served by an artificially created flow of water, usually engendered by a mill pond. It was the introduction of mill ponds which led to the earliest water right conflicts. The mill ponds raised the elevation of the river water, created an artificial head for a water wheel and, at the same time, incidentally established a pond. If the pond extended far enough upstream, it would slow the natural flow past one of the earliest mills, robbing its wheel of natural motive force. Similar disputes arose between latecomers whose mill ponds began to interfere with one another.

As a result of these early water right disagreements the riparian doctrine was born. In its earliest form, the doctrine stated that the owner of streamside (riparian) land was entitled to the continued natural flow of the stream. That approach, which came to be known as the *natural flow* version of the riparian doctrine, may still exist in humid portions of the United States. The natural flow version is, however, unsuited for a developed, industrialized economy. It may have assured nonconsumptive uses of water by riparian mill owners, but it was quite unsatisfactory for consumptive industrial or agricultural uses of water. The moment that riparian owners began to divert water from a stream and to *consume* part of it, as in irrigation or in the operation of thermoelectric plants, the natural flow version of the riparian doctrine caused them endless frustration. Their consumptive uses could never be secure from downstream challenges.

Consequently, the natural flow version was modified to become what is now known as the *reasonable use* version of the riparian doctrine. Under that approach, the riparian owner could change the natural flow of the stream if his use of the water was reasonable. Even the reasonable use version of the riparian doctrine, however, was useful only in those areas of the country where there was so much water in the surface streams that disputes over its use were rare. Otherwise, every dispute would end up in court and would hinge on the uncertain definition of reasonableness. In addition, even if the use was eventually found to be reasonable, the place of use was limited to riparian land, eliminating significant development away from the stream. Some riparian states, however, subsequently established administrative permit requirements for new uses and others have substantially modified the court-made law by legislation.

In those areas of the country where water was scarce, greater flexibility and certainty were essential and the riparian doctrine was replaced by a doctrine different in virtually every respect.

*The Appropriation Doctrine.* In the mid-nineteenth century, in the valleys and slopes of the high Sierra Nevadas and the Rocky Mountains, men and women seized with gold fever had an urgent need for water to work their claims. Unfortunately, however, there wasn't much water to be found. Although the mountain streams ran full during the spring snow melt, by midsummer stream flows were low. The riparian concepts did nothing more than encourage widespread violence over competition for water. Consequently, to promote peace and certainty, the early miners developed an approach to water rights which paralleled the law for establishing mining claims.

A mining claim was owned by the first person to stake it and work it. If water should be required and actually used to work that claim, the miner customarily was considered to have a water right. That water right was based on the appropriation or *taking* and *use* of the water.

When the first miner staked his claim and used water from a stream, he expected that no one else would subsequently "jump" his claim or interfere with his water. That expectation grew into a custom which was eventually formalized as today's prior appropriation doctrine.

The foundation of that doctrine is the concept of "first in time is first in right," simply meaning that the first person to take and to use water from a stream is entitled to continue his use—in spite of any subsequent demands for water from that stream system. If water still remains after that first use is made, then the second appropriator may use the water. Similarly, after the second use is made, the third user may then divert from whatever water may remain. As a result, on many western streams there is a list of priorities for the use of water. The earliest use of the water is given the first or most senior priority, while the most recent use of water is given the latest or most junior priority.

When the rights to use water are based on such priorities, it is said that the *prior* appropriation doctrine applies. That doctrine is found in most of the arid portions of the western states because, when there is too little water to satisfy all users, sharing of the remaining water would be of very little value to any user. In addition, where significant investments of capital and labor were required, the investor (even the lowly, scruffy miner) needed to have some guarantee that if water is essential to its value, the investment will be secure.

Not only did the appropriation doctrine secure investments, it also avoided several other problems created by the riparian doctrine. The early miners were trespassers on public lands. Consequently, they could not have been the owners of riparian land—as required by the riparian doctrine. In addition, they often wished to use the water wherever they found gold, even at nonriparian locations, far from the stream, a practice absolutely prohibited by the riparian doctrine.

While the appropriation doctrine has been subject to infinite refinements by the various western states, it remains essentially the approach of the early miners. Today's water rights held by miners, irrigators, municipalities, and industries all have been assigned priorities. For some stream systems, these priorities number in the thousands. Consequently, the administration of those priorities (mak-

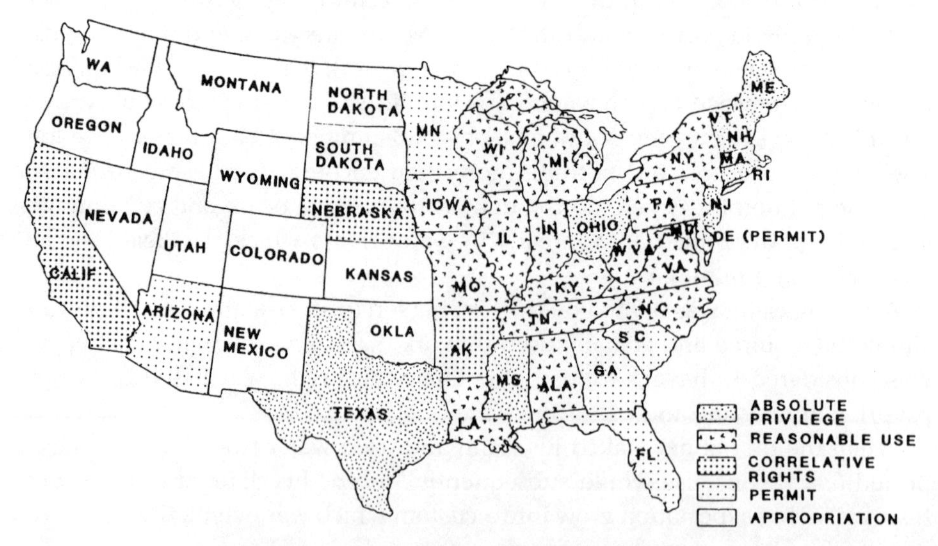


**FIGURE 11.** Predominate groundwater doctrines.

ing sure that each water user takes his proper turn) has developed into a major responsibility for state government. While riparian rights are usually enforced by individual lawsuits between competing users, appropriative rights are enforced by state water officials who ensure that each water user diverts water only when it is available to his priority. Only when state administration breaks down are the courts involved.

## *Groundwater*

Various approaches to the use of groundwater have been adopted by the states. These are roughly summarized in Figure 11.

The initial legal doctrines for groundwater were based on ignorance, compounded by traditional English concepts of land ownership. Since the Norman invasion, the English landowner has always been thought to own not only the surface of his land but also everything above it to the heavens and everything beneath it to the center of the earth. Consequently, when an Englishman first began to dig holes or wells in the earth, he was expecting to find water which he owned—simply because it was beneath his land.

*Absolute Privilege Doctrine.* In the earliest disputes between well owners, the courts confirmed the property owners' expectations. Judges concluded that, since the landowner owned everything beneath the surface of his land, including water, he was entitled to take it regardless of the consequences. The courts failed to recognize that the molecules of water did not, even then, remain stationary. Instead, when those molecules were withdrawn from the soil, they were eventually replaced by other molecules, often migrating from beneath a neighbor's land. When the migratory nature of groundwater was finally recognized, the traditional judicial concept of absolute ownership changed to a new approach, based on an absolute right or absolute privilege, to take whatever water (regardless of its source) that might be found beneath one's land.

Eventually, in the nineteenth century, American courts began to distinguish between types of groundwater, classifying it either as part of an underground stream having defined boundaries and direction of flow, or as percolating water, thought to ooze and seep through the earth without a defined rate and direction of flow. Courts decided that the same law which applied to surface stream water would also apply to underground streams. As a result, the absolute privilege doctrine was restricted to percolating water. That doctrine was satisfactory for virtually all purposes until mechanical pumps were mounted on wells.

*Reasonable Use Doctrine.* Around the turn of the century, large cities began to rely on wells in addition to the diversion or impoundment of major rivers—most of which had become so polluted that even our rugged forefathers found their water to be distasteful and unhealthy. One of these municipalities was the City of New York.

Near New York City was a truck farmer named Forbell. Because of the plentiful rainfall around New York, Mr. Forbell did not need to irrigate, except during very dry periods, but instead relied on natural rainfall and subirrigation from a high-water table. The City of New York, relying on the absolute privilege doctrine, decided to drill a number of large wells on a small parcel next to Mr. Forbell's farm. When large pumps were installed and the wells went into production, the water table dropped and Mr. Forbell's crops began to dry up. He sued the City of New York, seeking to prevent the city from continuing its pumping. The New York state courts were sympathetic and discarded the absolute privilege doctrine (under which New York City certainly had been within its rights) and adopted, instead, the reasonable use doctrine. Under that doctrine, as it was originally formulated, the overlying landowner could still pump all the water he could find *but* only so long as it was *used on the overlying land.* The overlying owner could transport the pumped water to a location *off* the overlying land only if none of his neighbors were injured by the pumping.

Subsequently, disputes began to arise over what was meant by the concept of overlying land. For example, cities such as New York would, by annexation, extend their boundaries into rural, agricultural areas. Sinking wells at the edge of city property, the municipalities would transport the water from the wells across town and maintain that it was still being used on the overlying land. The protection that the courts thought they were affording Mr. Forbell soon broke down and further refinements were made in many states. Even uses on the overlying land were unlawful if unreasonable, whatever that might mean.

While the concept of reasonable use did not lead to the certainty and stability necessary in the use of property, it has allowed the courts to make judgments on a case-by-case basis in order to protect landowners against the avarice of their neighbors. The approach works, however, only in those portions of the country where there is enough water so that disputes seldom arise.

*Correlative Rights.* A natural extension of the modern version of the reasonable use doctrine has been the correlative right doctrine, which is found in its purest form in the State of California. Under that doctrine, in times of shortage, scarce groundwater is shared by overlying landowners based on the extent of their overlying ownership. Such an approach may work when the utility of competing overlying uses is roughly equal. It does not work when an extreme use of large amounts of water is necessary for activities on a very small parcel of land. Consequently, even in California, the correlative right doctrine has been modified by a concept of prescriptive use, which is very similar to the doctrine of prior appropriation, under which the earlier user of water is entitled to continue that use at the expense of the Johnny-come-lately.

*Permit Systems.* The court-made law has reportedly been superseded by permit systems in states such as Delaware, Florida, Georgia, Hawaii, Minnesota, and South Dakota. Other states, such as Nebraska, New Jersey, New York, North Carolina, and Wisconsin, appear to have partial permit systems.

*Prior Appropriation Doctrine.* In many of the western states, concepts of consistency have dictated that even percolating groundwater be subject to the same legal doctrines as surface and underground streams. In other words, if a western state has adopted the prior appropriation doctrine for surface streams, it usually, *but not always*, has adopted the appropriation doctrine for percolating water as well. Under the prior appropriation doctrine, even wells are assigned priorities. Since withdrawals of hydraulically connected groundwater will eventually influence the flow of surface streams, the priorities of wells are usually integrated into the priorities of the diverters from surface streams. Because of the sometimes uncertain relationship between the withdrawal of water from a particular well and the diversion of water by a particular headgate, the interrelationship between groundwater and surface stream users has posed difficult and sometimes unsurmountable problems for state water administration officials. In fact, western water lawyers are fond of saying, "In a groundwater case, he who bears the burden of proof loses."

## Federal Water Right Doctrines

There are two federal doctrines which must be mentioned, the navigation servitude and federal reserved rights. The development of both will be treated in substantial detail in a subsequent chapter.

### *Federal Navigation Servitude*

Based on the constitutional power to regulate commerce among the states, the courts have recognized a navigation servitude, or easement, which favors the federal government. This means that any use of water from a navigable stream (even if sanctioned by state law) can be summarily interrupted, without compensation, should it interfere with navigation on that stream.

### *Federal Reserved Rights*

During the twentieth century, federal courts also fashioned another doctrine which gives water to the federal government, outside of any state allocation scheme. It provides that when the United States withdraws or reserves land from the public domain—for Indian reservations or national forests, for example—it also reserves enough water to ensure that the primary purpose of the land reservation is not entirely defeated. Such a reserved water right carries a priority as of the date of the land reservation. Water rights subsequently obtained pursuant to state water law doctrines are subordinated to the reserved right.

# RELATION OF ENGINEERING AND LAW

## General Considerations

In the exercise and administration of water rights under the appropriation and other doctrines, the fields of engineering and law come together. Simply stated, attorneys practicing water law must be able to relate the legal issues involved to the engineer, who is responsible for developing and explaining the factual situation, using available data and scientific and empirical methods. The engineer must be able to communicate clearly his factual conclusions to the attorney and other interested parties and explain how they were derived. Often this involves expert testimony in an administrative or judicial proceeding.

This process requires a high degree of communication between the engineer and attorney, and a mutual understanding by each of the other's discipline. While comprehension of essential legal principles is necessary, the engineer must avoid the temptation to draw legal conclusions. At the same time, the attorney must understand the limits of the engineer's scientific and empirical techniques used to develop factual conclusions, and not attempt to extrapolate results beyond the limits of the accuracy of the data and the formulas used by the engineer.

Both the law and the engineering applied to water rights change with time. The law changes through new legislation, court decisions, administrative interpretations, and in response to social, economic, and political conditions. The engineering techniques and methods used change with advances in technology, including adoption of research findings, availability of more sophisticated instrumentation, improvements in aerial photography and remote sensing techniques, accumulation of more complete and accurate data, and the lessons of experience.

## Sources of Law

One of the most difficult tasks for any lawyer is to outline, for an engineer, what is to be done in a water matter. The assignment must be made at the outset of a case, long before all the general and technical facts are known. Nevertheless, the final product of the engineer's work, generated months or even years later, must be relevant and admissible. The engineer also faces difficulties. He must maintain professional independence while organizing his work in a way that will efficiently meet the lawyer's needs.

One word of caution must be given at the outset. The lawyer must dictate the guidelines of the engineer's inquiry, the format of the engineer's work product, and, perhaps, the total budget for an engineering analysis, but the lawyer should neither dictate the substantive result of that inquiry nor selectively limit the information evaluated by the engineer in reaching a professional conclusion. To do so is to invite disaster in the courtroom when *voir dire* and cross examination eventually expose the folly.

To determine the scope of engineering work, the lawyer must first establish what is legally relevant in the situation at hand. He must collate the principles found in numerous legal sources—statutes, cases, and regulations. Occasionally, he may need to consider constitutional provisions as well. The most experienced engineer often finds this approach to formulating his scope of work unsettling. For those unfamiliar with the legal sources with which the lawyer must work, it can be downright infuriating. Nevertheless, when working within the constraints of our legal system, care must be taken to ensure that all the legal bases are covered and that extraneous matters (however interesting) be left to the academics.

This example may give the fledgling engineer an appreciation of the process. Assume that a Colorado water right owner now wishes to divert water for a residential subdivision, while his water right has been historically used solely for irrigation. Under the Colorado Constitution, the right to appropriate stream water can never be denied. Superficially there should be no impediment to the new appropriation. The knowledgeable lawyer and engineer, however, will immediately recognize that a new appropriation for subdivision purposes will have a priority so junior as to be virtually worthless. In order to enjoy a senior priority for the subdivision's water right, they will elect to change the use of the existing irrigation water right to that for a subdivision—while preserving its original senior priority. Colorado case law clearly states that the owner has an absolute right to change his existing irrigation water right. Both the cases and the statutes provide that any such change must not injure other water rights. The burden of showing lack of injury falls squarely on the person seeking to make the change.

The engineering analysis to show the lack of injury can be infinitely complicated. Since most clients do not have infinite resources, the engineering work often must be limited by the client's ability to pay. Consequently, the lawyer may ask the engineer to only evaluate the historic use of the existing irrigation water right—the amount, location and timing of historical diversions, depletions and returns. The lawyer may request the same information for the proposed subdivision use.

While the engineer may logically wish to make a more detailed investigation to ensure that the new subdivision use will not injure each of the other water rights on the stream, it may not be necessary. In a noncontroversial and straightforward change proceeding, it may be sufficient simply to compare the historic use with the proposed use and to show that the depletions of the new use do not exceed those of the historical use.

The scope and intensity of the engineer's work are directly related to the notoriety of the changes. If the change in question turns out to be controversial, then it may be necessary to drastically expand the scope and increase the intensity of the engineering investigation. For example, although the historical use under an irrigation right usually is limited to a 150–180 day irrigation season, water may be needed for the subdivision on a year-round basis. As a result, the engineer may need to fashion a storage plan to use excess historical irrigation depletions as augmentation for winter releases to compensate for subdivision depletions which occur outside of the irrigation season. Should the subdivision wish to take its water through wells, the result will be an even more complex engineering evaluation of how a stream diversion can be converted into a number of smaller well withdrawals, all without injuring other water rights. In the final evaluation, the engineer's work is circumscribed by the minimum requirements of the law, combined with the client's budget and the controversy spawned by that particular case.

After the lawyer and the engineer develop the engineering scope of work, another important issue remains to be addressed—whether the engineer's work is to be scrutinized by a court or an administrative agency. The world's best engineering work may be utterly useless if it is not admissible in evidence. Consequently, the lawyer and engineer must collaborate closely on how the engineer's work will be used. For example, in some states the engineer's work must be based only on facts which are already in evidence, requiring a laborious effort of developing exhibits showing the facts and numerical data upon which the engineer relied. In other states those underlying facts need not be in evidence (or even admissible) if they are of a type reasonably relied on by engineers when making the same sort of evaluation for out of court purposes. In either event, the engineer's work must be fashioned in such a way that it is ultimately admissible.

## WATER RESOURCE DEVELOPMENT

Water is one of our most vital, renewable resources, essential to the very existence of life. In addition to supplying daily sustenance to the animal and vegetable kingdoms, water generates power, serves as an artery of transportation, provides recreation, and has many other useful functions.

However, water can become a powerful destructive agent in the form of storms and floods. Uncontrolled water can destroy life and property, erode vast amounts of valuable soil, and lay waste to large areas.

As the population of the earth increases, accompanied by the continual urbanization of large population centers, the demands upon our water resources become even more severe. Therefore, it is important that we constantly strive to gain a better understanding of the occurrence, circulation, and behavior of water.

The development and control of water involves both the engineering and legal aspects of the right to divert and beneficially use water. Water resource development deals primarily with the land portion of the hydrologic cycle including precipitation, surface and subsurface flow, and evapotranspiration. Water rights often involve other disciplines such as agriculture, agronomy, aesthetics, forestry, fisheries, chemistry, and economics.

### Surface Water and Groundwater Development

The distribution of surface water as a resource presents many problems. Rainfall varies from region to region, and the amount of rainfall in any one place can vary widely from season to season and year to year. Even greater variations can occur in runoff and streamflows derived from precipitation and melting snow. The amount of water available for development in any one locality depends on the annual surface runoff, groundwater recharge, lake and reservoir storage capacities, and evaporation potential. Each of these factors can be measured and evaluated to determine the feasibility of water resource projects and also can be used to prepare development programs.

In some areas, groundwater is withdrawn to supplement or replace surface supplies. Shallow groundwater in alluvial flood plain deposits is generally considered a part of the surface stream system and is administered under the priority system. Deep groundwater may exist in aquifers that vary widely in thickness, extent, and physical characteristics. The evaluation of groundwater resources requires a knowledge of geology and hydrology and often involves the drilling and pumping of test wells to determine the aquifer's characteristics. The quality of surface water and groundwater is affected by geologic, hydrologic, and biologic factors which must often be considered in water resource planning.

## Environmental and Mineral Resource Analysis

Water is a primary environmental resource. A knowledge of its occurrence above and below the ground and its distribution in time and place is essential to understand the impact of proposed actions on the environment. Inventories of basic hydrologic data, evaluations of developmental effects on the water resources of an area, and the design of procedures and facilities to minimize adverse effects are essential features of environmental analysis. The water rights specialist is often part of a team involving all of the environmental and social sciences in a multidisciplinary approach to solving environmental problems.

Extraction and processing of mineral resources such as coal, oil, uranium, sand, and gravel are governed by a variety of federal regulations, state land use laws, and local zoning ordinances. In many cases, land must be properly zoned and federal, state, and county permits obtained before mining can start. This requires the operator to prepare detailed plans to protect the environment during the mining operation and to reclaim mined areas after the mining is completed. Mineral extraction may alter the surface water and groundwater systems, affect water quality, and disrupt water rights and irrigation ditches.

## Water Rights and Urban Systems

Water is a renewable natural resource. Nature furnishes a new supply at irregular intervals and in irregular amounts, as precipitation. Because of these irregularities in time and quantity, the demand for water may often exceed the available supply. Pollution can render water unfit for many uses. The right to use water is regulated and administered on a comprehensive basis by most states, particularly in the semiarid western states. State officials keep records of water use, receive applications for new water uses, and appoint water commissioners to supervise the distribution of water in accordance with established water rights. Court actions are often required to adjudicate new surface and groundwater rights and to convert existing rights to new uses. In some states, water rights can be sold or transferred like real estate and may be bought and sold separately from land titles.

The continuing expansion of metropolitan areas involves the conversion of rural land areas to buildings and pavements. In many cases, the replacement of open land with asphalt and concrete causes significant changes in the hydrologic character of the area. These changes include increases in peak runoff rates, degradation of stream water quality, and the alteration of water and sediment distribution patterns. Providing water for people involves recognition of how the water resources system interacts with other urban features, such as transportation, solid waste, open space, and building development. Solving water-related urban problems also involves close cooperation with specialists in law, finance, engineering, community development, and city planning.

# 2

# GENERAL PRINCIPLES

## FUNDAMENTAL CONCEPTS OF THE APPROPRIATION DOCTRINE

In the western states, the right to use water is usually established by the prior appropriation doctrine. That doctrine is characterized by the expression "first in time is first in right," which simply means that the first person to use water is thereafter entitled to the first opportunity to use the same amount of water for the same purpose.

There are two major exceptions to the general rule of appropriation in the western United States. The first exception is found in those states which have combined the riparian and the appropriation doctrines. The second exception, found even in some states which have totally rejected the riparian doctrine, concerns the type of water to which the appropriation doctrine will be applied. Although the waters of surface streams may be universally subject to the appropriation doctrine, groundwater and diffuse surface water often are not.

### Appropriation Systems

While the first in time expression is generally correct, there are many qualifications. Only in Colorado does the pure appropriation doctrine, under which a person obtains a water right simply by *using* water, still exist. In most other western jurisdictions, after years of legislative tinkering, one gets a water right only by application to an administrative agency. In those states, he who files first is "first in time" and obtains the best water right.

## *Mandate System*

Those few remaining states where a water right can be established without administrative action are called mandate states. The individual right to make an appropriation is mandated by the state's constitution. Only Colorado still falls squarely in that category—and then only for the surface waters associated with a natural stream.

In a jurisdiction retaining vestiges of the mandate system, a water right is created by the physical action of the appropriator—the person who takes the water and uses it. While the water right may later be confirmed by judicial or administrative action, until then there is usually no official, written record of its existence. While such an unrecorded water right may not enjoy full protection of state water rights administration, it lurks in the shadows to bushwhack the unsuspecting lawyer or engineer.

## *Permit System*

Toward the end of the nineteenth century, an engineer familiar with the vagaries of Colorado's mandate system became Wyoming's first state engineer. Elwood Mead, who later would head what is now the U.S. Bureau of Reclamation, was determined that Wyoming would not repeat Colorado's mistakes. He devised an approach to water rights which is still in use in Wyoming, and which has been adopted, in various forms, in most of the other western states.

The existence of unrecorded water rights offended Mead's engineering sense of order. Following statehood in 1890, water rights in Wyoming fell under a strictly enforced permit system. After making provision for territorial rights which were created under the mandate system prior to statehood, Mead's system required that all new appropriations be applied for and approved by the state engineer. Should the state engineer find that unappropriated water was available, he would issue a permit to appropriate. Armed with a permit, the applicant would then make an appropriation.

The permit was effective for a fixed period, although the state engineer could grant extensions. If no appropriation was made during the life of the permit, it would lapse. If an appropriation conformed to the permit's terms, the appropriator would then apply to the state Board of Control (the state engineer and his four division superintendents) for a certificate of appropriation.

While the Wyoming permit system, as well as its numerous progeny, have been plagued with many difficulties, the approach has worked remarkably well, especially when adequately funded by state legislatures and when the state engineer has been both discreet and aggressive.

## Beneficial Use Is the Measure

In most of the western states, the term "appropriation" is defined as being the diversion of water and its application to a beneficial use. Even in those permit states where a water right is initiated by application to an administrative agency, a beneficial use must be made of the water before the right is perfected or finalized. In subsequent dealings with the water right, in addition to its relative priority, the most important attribute of that right is its beneficial use. Normally, a water right obtained for one beneficial use cannot be used for another beneficial use without some administrative or judicial approval. What constitutes a beneficial use is a question of state law. Historically, most states recognize beneficial use as being virtually all utilitarian uses, including irrigation, mining, milling, stockwatering, and domestic consumption. Recently, nonutilitarian uses such as recreation, fish propagation, and aesthetics, have achieved some acceptance as beneficial uses.

In addition to its type of beneficial use, each water right is also evaluated in terms of the intensity of its use. Intensity of use has two aspects, the amount and timing of water actually taken for that beneficial use as well as that water consumed by the beneficial use. Among water engineers and water lawyers, that intensity of use is usually referred to as "historic use."

## Water Right Is an Interest in Real Property

In virtually every state, a water right is considered to be an interest in real property, just as is the ownership of land. The western states differ greatly on whether or not a water right may be conveyed separately from the land which it serves.

## Initiation and Perfection of the Right

Although previously described, it may be useful to summarize the way in which water rights are initiated in the western states.

### *Appropriation*

Traditionally, water rights were created by the physical act of taking water from its source and applying it to a beneficial use, as in a pure mandate state. Since the turn of the century, numerous refinements have been made. In most states a permit system has developed under which a water right is not created by individual action, but by an administrative official granting an application for a permit to appropriate.

Even in a mandate jurisdiction, such as Colorado, appropriations no longer must include the actual application of water to a beneficial use. Because of the needs of growing municipalities, a "conditional water right" was created. Although municipalities needed to secure water supplies for future expansion, they were often in no position to make a traditional appropriation. It was impossible to apply water to a beneficial use when their inhabitants already had all the water they needed. On the other hand, because all of the conveniently available water was rapidly being appropriated, the municipalities desperately needed to establish and preserve priority dates for specific amounts of water for future use. Not wanting to force cities and towns to remain inactive while unappropriated water was available and then be forced later to acquire water when it was actually needed (perhaps by condemnation), in the early twentieth century the courts and legislatures created what is now known as a conditional water right.

To obtain such a right, an appropriation is made by the coexistence of two elements: (1) the subjective intent to take a specific amount of water for an identified future use and (2) physical acts (usually a survey of a headgate or dam location) which demonstrate the intent, which put other water users on notice that an appropriation is being made, and which are substantial steps in completing the diversion or impoundment. Once created, a conditional water right must be carried through to completion with "due diligence"—determined by the courts on a case by case basis, by asking whether the appropriation has been prosecuted with the zeal of a reasonable man under all the circumstances. Since conditional water rights may stay alive for decades, the Colorado legislature has now provided that, in the absence of quadrennial court findings of diligence, the right will be cancelled.

### *Adjudication*

Water rights are adjudicated by judicial or administrative action, depending on the state. In states with procedures based on the mandate system, the water right is adjudicated (officially confirmed and awarded a priority) by a court decree. Indeed, in Colorado, a discrete system of specialized water courts has been established for precisely that task. Once a water right (either conditional or absolute) is adjudicated, the court's decree becomes the warrant of the state engineer for the administration of that and other adjudicated rights.

In permit jurisdictions, adjudication is typically an administrative rather than a judicial function. Permits to appropriate evolve into perfected rights (usually, but not always, evidenced by certificates) by the quasi-judicial action of an administrative entity, such as Wyoming's Board of Control.

In either type of jurisdiction, once a water right has been adjudicated, nothing more must be done to keep the water right alive, except to use it.

## Administration of the Right

Water rights are usually administered by a state agency, typically headed by the state engineer. The agency's primary responsibility is to develop lists of water rights on each stream or source, arranged in order of priority—the water right which was first in time is first on the list, that which was last in time is last on the list.

Through a network of field representatives, often called water commissioners, the state engineer ensures that the water to which seniors are entitled is not taken by juniors. This is a task which is very simple to describe but quite difficult to carry out.

## *System of Calls*

The water rights on *most* western streams are virtually self-administrating. Familiar with one another's priorities, long-time neighbors voluntarily restrict their water usage to maintain the priority system. As stream systems get larger and the need for water becomes more extreme (such as during a drought), voluntary compliance with the priority system often breaks down. When that happens a system of "calls" is triggered. A senior water right owner, who feels that he is not getting the water to which he is entitled, will call up the water commissioner and ask him to stop diversions by any juniors until the senior's water right is satisfied. By doing so, the senior water right owner is "putting a call" on the river. If that senior water right owner had an 1891 priority, it would be said that an 1891 call was being put on the river. Upon receipt of the call (which in many states must be in writing), the commissioner then notifies junior water right owners to stop their diversions. Usually, the commissioner can enforce a call merely by a series of telephone conversations with people he has known all his life.

Unfortunately, there has been an increasing inclination to ignore the commissioner's telephonic requests. Consequently, the water commissioner is often required to "post the headgate." That simply means that he goes to the diversion facility of a ditch and tacks up a notice (often an administrative order) stating that the ditch is to be shut off until further notice. If voluntary compliance with the priority system has deteriorated to this point, the ditch owner usually tears up the posted notice. When that happens, it is not long until the headgate operations are supervised (usually with a padlock) by the water commissioner.

Shortly thereafter, the padlock is blown apart with a large-caliber handgun or rifle. Within a few days, a court will issue an injunction against further diversions through that headgate, along with the admonition that if the injunction is violated, the ditch owner will go to jail. Usually, when that happens, no further diversions are made through the ditch—except at night.

As long as an atmosphere of good will and voluntary compliance exists on a stream, administration of water rights is not terribly complicated or difficult. Once voluntary compliance breaks down, however, pandemonium ensues and state administrative officials are faced with an impossible task. Nowhere is that task more difficult than in the conjunctive use of groundwater rights and surface water rights. Groundwater withdrawals made from the river alluvium or from other sources in hydraulic connection with the river will eventually affect the amount of water available for diversion by senior ditch owners. To call out junior wells at just the right time (an anticipatory call) requires the wisdom of Solomon and the nerve of a dance hall gambler.

### *Rules and Regulations*

When the administrative system becomes totally chaotic and the state engineer is no longer able to cope with a multitude of individual enforcement actions, he will often promulgate rules and regulations for the use of water from a particular source. Those rules and regulations usually curtail all diversions through ditches and storage in reservoirs at such times as may be necessary to allow water to remain in the stream to meet interstate compacts. In addition, those regulations usually prohibit all withdrawals of water from wells until each well owner can establish that his withdrawal will not deprive any senior water right owner of water to which he is entitled. The courts usually uphold those regulations.

### *Futile Calls*

The futile call doctrine is one exception to the strictly enforced priority system. Under that doctrine, a junior right may not be called out if the water that is diverted would not otherwise reach the headgate of a senior water right. Such a determination is fraught with uncertainty. In many jurisdictions it has become administrative practice to assume that curtailment of a small diversion (one of 1 cfs or less, for example) will not significantly increase the supply of water available to downstream seniors. As water administration becomes tighter, rules of thumb are abandoned and eventually junior diversions will be shut down if any measurable amount of water will thereby reach downstream seniors.

### *Reasonable Means of Diversion*

Under the common law, each diverter was expected to have a reasonable means of diversion. The word reasonable has proved to be a point of contention for lawyers and courts. Most definitions of this concept state that a diverter from a stream may not command the entire flow of the stream simply to divert a small portion of it. The United States Supreme Court's opinion in the *Schodde* case is the most famous application of this doctrine.

The *Schodde* plaintiff diverted water from Idaho's Snake River through water wheels. The defendant constructed a downstream dam which severely raised the stream level leaving the water wheels ineffective. Seeking damages, the plaintiff lost when the court reasoned that an appropriator may not control the entire flow of the river in order to divert only a small portion of that flow. Following the announcement of *Schodde,* a number of states have enacted statutory versions of the same doctrine. The reasonable means of diversion requirement is now universally referred to as the Water Wheel Doctrine.

Until recently, it was thought that the Water Wheel Doctrine applied only to diversions from streams. A 1984 opinion of the Colorado Supreme Court, however, extended the doctrine to wells. In a case arising out of litigation over state engineer's rules and regulations for Colorado's San Luis Valley (Rio Grande), it was successfully argued that, because of the interrelationship between the water diverted by headgates from streams and the water withdrawn by wells from hydraulically connected underground sources, it was unreasonable to stop well withdrawals simply to leave water in the stream for surface stream diverters. The court ruled that under certain circumstances, surface diverters must drill their own well before calling out junior well owners taking water, albeit at a much lower depth, from the same source.

### *Preferences*

Many state constitutions and statutes contain a preference provision which purports to give preference to certain uses (such as domestic and stock watering) in time of shortage, regardless of their relative priority. Few states honor this provision. In Colorado, for example, the preference provision simply gives a preferred user the power to condemn a water supply from a senior priority. In many permit states the preference provision is used by the state engineer to help him choose between competing applications for permits to appropriate. Since each state copes with its preference provision differently, the only hard and fast rule is to not take them at face value—a useful approach when evaluating any water concept.

## Loss of Water Rights

Water rights may be lost by adverse possession, waiver, estoppel, and laches, but the two most significant ways in which they are lost are abandonment and forfeiture. Both concepts are founded on an age-old precept: "use it or lose it."

### *Forfeiture*

Most western states have forfeiture statutes, under which non-use of a water right for a specified period of time (typically five years) means that the water right is lost. It simply disappears, absent any extenuating circumstances which may be provided by statute. While forfeiture often occurs only after administrative or judicial action, forfeiture laws are usually strictly enforced.

### *Abandonment*

Abandonment, unlike forfeiture, usually is not a statutory provision. It has its roots in the common law applicable to all forms of real property. Under the common law doctrine of abandonment, when one ceases to use property *and* intends never to use it again, the property is considered to be abandoned. Consequently, abandonment can be triggered almost instantaneously, without waiting for years of non-use. For example, if one intended never to use a water right again and, in fact, did not use that water right, abandonment could take place in a matter of seconds. In most abandonment cases, the difficult aspect is the owner's intent. Regardless of what their intent might have been, few owners will readily admit in court that they intended never to use their water rights again. The law has developed a presumption under which non-use of a water right for an extended period of time will create a presumption that its owner intended to abandon it. The courts generally are loathe to abandon water rights and usually allow the presumption to be rebutted by the flimsiest evidence.

### Acquisition and Disposition of Existing Water Rights

Water rights are bought, sold, and leased in a fashion very similar to land. The typical conveyance of water rights is by deed. The details of these types of transactions are found in the section called "Acquisition and Disposition of a Water Right," later in this chapter.

### Adaptation of Existing Water Rights to New Needs

When a community expands or a new industry decides to locate in one of the western states, it usually must acquire water rights. In many of the western areas the streams are overappropriated, meaning that all the water in those streams has already been appropriated, during at least some part of the year. When one acquires a water right, he also acquires the right to change it adhering to the requirement that the change may not injure any other water right—especially junior water rights. Consequently, the municipality or the industry may change what probably was originally an irrigation water right to a new use.

## ELEMENTS OF A WATER RIGHT

In the western United States most water rights are evidenced by an official piece of paper, be it a decree, permit, certificate, license, etc. That paper contains most of the information needed to effectively describe a water right. The document is usually filed with the agency which issued it, such as a Colorado water court or a state engineer's office. The following information can be found on the document: the name of the water right, the location of its point of diversion, the source from which it takes water, the use to which the water it takes may be put, the amount of water it may divert, and its priority date.

### The Water Right Document

These organic documents contain the basic information required to begin an investigation of a water right. Examples of a decree of a Colorado water court, a permit issued by the State Engineer of Wyoming, and a certificate issued by the Wyoming Board of Control can all be found in "Perfection of a Water Right," later in this chapter.

#### Name

Water rights rarely have a truly distinctive names. Most water rights bear the name of the original appropriator or a descriptive name based on their geographical location. Consequently, just as there are hundreds of Joneses in the telephone directory, there are also hundreds of Jones' ditches in the western United States. Similarly, there are innumerable ditches with the same geographical name such as the Southside Ditch. The name is only a beginning in describing a water right, since there are often more than one water right sharing the same name and since the same name often refers both to the water right as well as to its diversion structure.

A water right is a different interest in real property than its diversion or distribution facility. Ownership of the water right simply gives one the right to use water. That ownership interest is usually separate and distinct from the ownership of a ditch and its headgate, or a dam and a reservoir site. Normally, both the water right and its physical facilities are held in common ownership. Frequently, especially for large capacity ditches or reservoirs, a large number of water rights, all with differing ownership and priorities, may be exercised through the same structure.

## Location

Most water right documents describe the location of the structure or facility through which the water right is utilized. Frequently, for example, in the case of a ditch, the legal description of its headgate location can be found in the water right document. The use of that location, when describing a water right, avoids misunderstandings which arise when two different water rights have the same name. Although recently perfected water rights usually are described with careful attention to legal descriptions, the very oldest (and most valuable) water rights often have descriptions which defy precise location, such as "20 yards south of a 6 inch aspen tree with the initials V.T. carved on it." For a water right so described and adjudicated in 1880, odds are that the aspen tree no longer exists. In addition, legal descriptions are often internally inconsistent. For example, a headgate described as being "on the north bank of La Jara Creek in the W/2 of the NW/4 of Section 8" is particularly confounding if, by examination of a map, you discover that La Jara Creek does not now and never has flowed through Section 8.

Another important location is often omitted from the basic water right document—the location of the land served. The vast majority of water rights in the western United States are for irrigation purposes. They were appropriated to serve a particular parcel of land. On the forms later in this chapter, you will notice that the Colorado decree, the Wyoming permit, and the Wyoming certificate all describe those lands. If this information is missing, other documents, such as court testimony or exhibits, may establish the identity of the land served by a particular water right.

## Source

Virtually all basic water right documents describe the source from which the water right is entitled to take water. As with structure names, there is a substantial amount of duplication in any given state. It is probably safe to wager that most western states have several Horse Creeks. In addition, especially with respect to groundwater, the description of the source is only vaguely helpful, for instance, "Arapahoe Formation" or "underground water tributary to the Conejos River," which is an interesting description since the well in question may be 50 miles from the Conejos River.

Separately, each of the elements described above (name, location, and source) may be insufficient to positively identify or describe a water right, but together those elements usually provide a degree of certainty which can be relied on by a lawyer or engineer.

## Use

The water right document typically describes the beneficial use for which the water right was appropriated, such as irrigation, stockwatering, domestic, or municipal. It is important to realize that the document's recitation of use is only a starting point. For example, if a water right was used for flood irrigation, as both a practical and theoretical matter, it may be limited to flood irrigation in the future. Under flood irrigation, there were probably significant returns of unused water to the stream from which it was originally diverted. Should the water right owner change to more efficient sprinkler irrigation, the returns may be severely diminished, thereby extending the use of the water right or impermissibly increasing the burden which exercise of the water right imposes on the stream.

## Amount

The amount for which a water right is adjudicated is usually expressed in cubic feet per second (cfs), acre-feet (af), or gallons per minute (gpm).

Water rights based on ditch diversions are usually described in cfs, inches, or miner's inches. Inches or miner's inches can be converted to cfs, based on the statutes or customs of a particular state. As a rough rule of thumb, 40 inches is 1 cfs. Well withdrawals are often measured in gallons per minute, with approximately 450 gpm equalling 1 cfs. Both ditch diversions and well withdrawals are sometimes described with an annual volumetric limitation, measured in acre-feet; 1 cfs diverted for 24 hours is approximately 2 af.

Storage water rights, the impoundment of water in reservoirs for a later use, are typically measured in acre-feet. Once in a great while, a storage right may be described in cubic feet or in gallons.

## Priority

Perhaps the most important piece of information to derive from a water right document is the relative priority of the water right. However, the priority may be the most difficult to determine. Traditionally, the priority was the first date on which water was actually taken and applied to a beneficial use. As previously described, under the various processes of adjudication, that date may no longer establish the water right's priority. In Colorado, the appropriation date is important in determining a water right's priority only in relation to another water right which was perfected in the same adjudication. In other words, if water rights were adjudicated at different times, the water right with the earliest adjudication date is usually the senior water right. In a permit state, such as Wyoming, the date the application for permit was filed is most important, notwithstanding the date upon which the application was granted and the permit issued or the date upon which the water right received a certificate from the Board of Control.

While important dates can be discovered only by referring to the basic water right document, the relative priority of a water right is usually determined by reviewing the basic water right documents for all other water rights having the same source. Fortunately, that tedious job has already been done, since most of the state engineers maintain a list of priorities of the various water rights on each source of water.

## KINDS OF WATER RIGHTS

Within any given state, there are usually several kinds of water rights, each treated somewhat differently and each requiring special attention. In a work of this nature, only the general types can be discussed. In dealing with a specific state, one must master the nuances applicable to that particular state's water rights.

### Direct Flow

Water rights in the West often are associated with ditches and canals. These are known as direct flow rights, usually measured in cfs or in inches. However, direct flow rights include any water right in which water is appropriated for *immediate* application to beneficial use. Consequently, water withdrawn through a well (usually measured in gpm) would also qualify as a direct flow water right.

### Storage

Another common water right, associated with a reservoir, is a storage right. A storage right is one in which the water is appropriated not for immediate use but for subsequent use at a later date. Questions sometimes arise over whether a particular water right is a direct flow or a storage right. This usually occurs when irrigation water is diverted through a ditch into a small holding pond where it is collected until a sufficient irrigating head is developed, at which time water is released for irrigation. By way of a very rough rule of thumb, water which is held more than 24 to 48 hours is often considered to be the subject of a storage right rather than of a direct flow right.

Storage rights generally are far more junior than direct flow rights. The better direct flow rights were established during the last century when unappropriated water was still plentiful during the irrigation season. Once all the water available during the irrigation season had been appropriated, it became necessary for subsequent irrigators to construct upstream reservoirs. These reservoirs collected the excess spring runoff, holding it until the irrigation season when it could then be released into the stream for downstream irrigation diversions, usually by shareholders or persons having a contractual relationship with the company which constructed and held the storage right to impound water behind the dam.

An almost universal requirement throughout the West is the one-filling rule for storage rights. This means that once a storage right has been satisfied, it cannot thereafter (during the same year) be the basis of additional storage. In other words, once a reservoir is filled under its storage priority and emptied to satisfy downstream irrigation requirements, the same water right may not be the basis of the refilling of a reservoir during that year. As a result, each storage right is limited to one filling per year. The actual administration of this rule varies from state to state and is not always consistent within a particular state.

In some states, a second system of storage water rights exists, usually referred to as stock tanks, which are small reservoirs on intermittent streams. This system is based on the assumption that the waters of those streams make an insignificant contribution to perennial streams and rivers. Consequently, each intermittent stream may be considered a virtually self-contained entity for which an independent system of priorities is appropriate. In the event that an intermittent stream does make a material contribution to a perennial stream, its system of stock tank priorities is usually subordinated to the priorities established on the perennial stream.

## Perfected Water Rights

Theoretically, a water right is perfected only after nothing more need be done from an administrative or judicial standpoint with respect to that water right. As described previously, it is necessary to exercise that perfected water right or it may be lost through abandonment or forfeiture.

It is important to understand that in most states unperfected water rights can be exercised just as if they were perfected. As a result, during the period before a conditional water right becomes absolute or a permit is certificated, diversions or storage can be made under the incompletely perfected water right. On the other hand, in many states, should the unperfected right not be perfected within the allowed statutory period, then it is possible that the right will be lost even if it has actually been exercised.

For a more comprehensive treatment of the perfection process, see "Perfection of a Water Right," later in this chapter.

## Enlargements and Extensions

As development progressed in the West, it became clear that some ditches and reservoirs (and their water rights) were inadequate to serve all lands beneath them which could be served by gravity flow. Consequently, if unappropriated water remained available, an additional water right would be claimed for the ditch or reservoir. In some cases, the ditch or dam was already large enough to handle more water but lacked the perfection of the water right enabling the owner to enlarge his diversion or storage. In most cases, however, the ditch or reservoir had to be enlarged or the ditch, perhaps, extended to serve more land. Subsequent water rights in the same structure are usually referred to as enlargements or extensions. It is not at all unusual to find a number of water rights, with varying priorities, associated with the same structure. For example, one could easily find, associated with a *structure* called the Southside Ditch, *water rights* bearing the names of Southside Ditch, Southside Ditch, First Enlargement, Southside Ditch, Second Enlargement, and so forth. With respect to reservoirs, it is commonplace to see the concept of extensions and enlargements used to defeat or circumvent the one-filling rule.

## Shares

One of the first cooperative efforts concerning water rights in the West was the development of mutual ditch companies. Members pooled their resources, constructed facilities, and obtained water rights, in all of which members shared based on their stock ownership. In most states, the owners of shares in a mutual company are considered to be the actual, or at least the beneficial, owners of the facilities and the associated water rights. As a result, subject to restrictions in the articles or by-laws of the company, the shareholder usually can transfer his shares (his pro rata interest in the water right and his pro rata interest in the facilities) to others, in much the same way he could transfer the ownership were he the exclusive owner.

## Instream Flows

In many of the western states, it is possible to obtain water rights for in-place uses, instream flows or lake levels, in addition to more utilitarian beneficial uses. The right to do so is usually conferred by statute and may be limited to state agencies. The purpose of such water rights is to preserve fisheries as well as aesthetic and recreational features of a body of water. If a stream is involved, the water right usually consists of a constant flow, measured in cfs within a specific stream reach, and varying perhaps by the month or day. Lake levels are similarly described either by elevation above mean sea level or af. Although most water rights for in-place uses are quite junior in priority, they nevertheless pose special problems for other water rights. When those other rights are the subject of change proceedings, it is usually quite difficult to avoid the proscribed injury to in-place uses. Consequently, even junior water rights for instream flows effectively freeze development on a stream, often eliminating the possibility of future changes of senior water rights.

### Developed and Imported Water

The most valuable water right in the West is one which is totally consumptive. This usually occurs when water is transported completely away from its original source, any water not consumed being returned to another water source not connected to the original. When that happens, the water is referred to as developed or imported water—at least with respect to the receiving source. An example would be water diverted on the west side of the Continental Divide but delivered and used on the east side. The same situation occurs when groundwater from a completely confined aquifer is brought to a surface stream.

In either event, the importer or developer of the water has free rein for its use in the receiving basin. How the water is used is of no concern to other users from the original source. Similarly, users from the receiving source cannot complain of the developer's use (absent some contractual arrangement) since the water would not be there under natural conditions. Consequently, the importer may reuse and make successive uses to the extinction of the developed water in a manner independent of priorities on the receiving source.

## PERFECTION OF A WATER RIGHT

The process for perfection of a water right vary significantly from state to state. A description of that process is illustrated with two states: Colorado (where the mandate system is largely still applicable) and Wyoming (the earliest of the permit states).

### Judicial Action—The Colorado Example

Perfection of a water right in Colorado requires that a court decree be obtained. Based on major watershed boundaries, the state is divided into seven water divisions, each having its own water court for the determination of water rights and its own division engineer (working under the state engineer) for the administration of rights.

Although it is fashionable to say that all of the waters in Colorado have been appropriated, during certain parts of the year, a number of streams still yield unappropriated water. While much of the water court's business is concerned with the changes of water rights, a large number of appropriations still are perfected in the water court.

The perfection or adjudication process now existing in Colorado was established in 1969, in an effort to simplify the process and make the courts accessible to those persons not wishing to go to the expense of engaging a lawyer. By and large, this system has worked well. The majority of applications are disposed of without adversary courtroom proceedings constrained by formal rules of evidence. The process is as follows:

1. At any time any person may file an application for the determination of a water right. That determination is a decree issued by the water court which confirms the existence of the water right and assigns it a priority.
2. When the water court receives the application, the water judge refers the application to a water referee, who is to make an informal investigation and, perhaps, conduct hearings on the application.
3. In the month following the filing of the application, the court clerk publishes a court "resume," which summarizes all of the applications which have been filed with that particular water court during the preceding month. The resume appears in newspapers of general circulation and is mailed to all persons on the clerk's mailing list. In essence, a resume is a newsletter showing what new activity has occurred in the water rights business during the preceding month.
4. By the end of the second month, any person who wishes to oppose the application must file a statement of opposition, setting forth the legal and factual basis for their opposition.
5. At the end of the statements of opposition period, the water referee either:
   a. decides that the application is so controversial that it should be rereferred to the water judge, or
   b. informally investigates the truth of the matters set forth in the application.
6. Part of the referee's investigation involves a conference with the division engineer, who points out problems and supplies information pertinent to the investigation.
7. At the conclusion of this investigation, the referee makes a ruling and submits it to the water judge.
8. Within 20 days of the ruling, any person may file a protest to that ruling.
9. If no protest is filed, the water judge typically endorses the referee's ruling, making it the decree of the court.
10. If, however, a protest is filed, the water court conducts a hearing de novo on the application and all the rules of civil procedure and evidence are applicable.

Salient documents in a very simple Colorado water right application appear after this discussion. This application was for a small, uncontested water right (the Young Sump) near Fort Collins, Colorado. When reviewing the pages, note the following attributes of each document:

1. The application was filed on December 31, 1981, in the Division 1 Water Court. The last day of the year is the busiest day in the water court. Under the existing adjudication process, water rights adjudicated based on applications in one year are all senior to water rights adjudicated based on applications filed in subsequent years. Consequently, there is no advantage to filing an application early in the year and all sorts of tactical and strategic advantages in filing an application late in the year for which the resume is published in the following year. This application was filed by a lawyer assisted by an engineer, although there is no reason why the applicants could not have filed for themselves. Because of complex factual background, the applicants wisely selected an experienced water lawyer and engineer. Usually, the applicants may simply go to the Water Court, obtain a standard form, and fill out their own application.
2. Following the application is an excerpt from the December 1981, resume distributed in January of 1982. A succinct summary of the application appears below the court-assigned case number (81-CW-459). In Colorado, the resume publication is assuming increased importance since the jurisdiction of the water court is limited to those matters which appear in the resume. If the resume incorrectly describes the application in a substantial way, the water court is without jurisdiction to grant the application.
3. Following the resume is a summary of the consultation held between the division engineer and the water referee. The division engineer recommended the application be approved but expressed concern about the extent of historic use under this right and made it clear that this right was considered to be a very junior one.
4. Based on the comments of the division engineer, the water referee asked the applicants' lawyer to provide further information concerning the historic use, particularly the priority date of the appropriation for which the application was made. The lawyer submitted two affidavits:
   a. An affidavit by Jamia Riehl, in which she testified that she had used the water of the sump to irrigate from 1962 until 1971, when the underlying property and the water right were sold to the applicants.
   b. An affidavit by Paul Waag testifying to his knowledge of the use of the water through the sump, dating back to 1931.
5. Following the affidavits is the referee's ruling for the Young sump in which the water right is described. At the bottom of the second page there is a form provided for the water judge to sign the ruling and to make it a decree of the court, in the event that no protests are filed.
6. Since no protests were filed, the water right received its decree on August 29, 1983, some year and a half after the application was filed.

This process is typical of the majority of water rights adjudicated in the Colorado water courts. Most go through the procedure without opposition but with careful scrutiny by the water referee. The water referee awarded a 1930 priority date to the Young Sump, rather than the 1894 priority date which was sought in the application. Undoubtedly, he awarded the later date simply because the affidavits submitted by the applicants established the use of water beginning only in the early 1930s. While the affidavit of Paul Waag makes it clear that the water right had been used prior to that time, there was no hard evidence based on personal knowledge of any affiant or witness to that fact. Consequently, the referee viewed the evidence strictly and quite properly awarded a priority date based strictly on the evidence rather than on surmise and conjecture.

In a small minority of the cases filed with the water courts, there is substantial controversy and extensive hearings before the referee and the court. Resolution of such a dispute can require 12 to 15 weeks of trial. While complex litigation is unavoidable when dealing with a scarce resource such as water, the Colorado adjudication system has achieved its intended purpose of eliminating needless controversy and expense.

FILED IN WATER COURT
DIV. I
STATE OF COLO.

81 DEC 31 PM 3:45

OFFICE OF
WATER CLERK

IN THE WATER COURT IN AND FOR

WATER DIVISION NO. 1

STATE OF COLORADO

Case No. 81CW459

| | | |
|---|---|---|
| CONCERNING THE APPLICATION FOR WATER RIGHTS OF ROBERT A. AND LYNN YOUNG IN LARIMER COUNTY, COLORADO. | ) | APPLICATION FOR WATER RIGHT IN SURFACE AND UNDERGROUND WATER |
| TRIBUTARY INVOLVED: SPRING CREEK, TRIBUTARY TO THE CACHE LA POUDRE RIVER | ) | |

1. Name, address, and telephone number of applicant Robert A. and Lynn Young:

   c/o David F. Jankowski
   Nossaman, Krueger & Marsh
   Kittredge Bldg.
   511 16th Street, Suite 300
   Denver, CO 80202

   Applicant's residence address:

   1601 Sheely Drive
   Fort Collins, CO 80526
   (303) 493-0293

2. Name of structure:

   Young Sump

3. Legal description of point of diversion:

   The Young Sump is located in the NW1/4, NW1/4, Sec. 23, T 7 N, R 69 W, 6th P.M., 670 feet East of the West line of said section and 515 feet South of the North line of said section. The street address of the point of diversion is 1601 Sheely Drive, Fort Collins, Colorado.

4. Source:

   The Young sump is supplied by two sources of water. Each source has a surface and ground water component. The first source is underground and surface water occurring naturally in an unnamed slough, located in Sections 22 and 23, T 7 S, R 69 W, 6th P.M., Larimer County, Colorado. Said source supplies the Young Sump directly. The second source is surface and underground water collected in the Tombough Drain and the B.B. Harris Drain. Said drains are also located in sections 22 and 23, T 7 S, R 69 W, 6th P.M., Larimer County, Colorado. Said drains are more specifically described as follows:

81CW459

Tombough Drain:

The head of the Tombough Drain is located at a point 240 feet south of the Northwest corner of Sec. 22, T 7 N, R 69 W. The ditch follows a line bearing N 81° 05' E 560 feet thence S 24° 30' E for 800 feet thence S 72° 15' E for 384 feet thence N 84° 10' E for 810 feet thence N 83° 30' E for 840 feet thence N 88° 05' E for 560 feet thence N 65° E for 252 feet where the end point of Tombough drain is the beginning of the B.B. Harris Drain.

B.B. Harris Drain:

The B.B. Harris Drain follows a line bearing from the point of beginning North 60 feet thence N 76° E for 110 feet thence S 83° E for 775 feet thence S 90° E for 770 feet thence north 100 feet thence East for 60 feet thence North for 180 feet thence S 76° 30' E for 820 feet thence S 23° 30' E for 270 feet to the end point of Tombough and B.B. Harris Drain.

Each of said drains collects surface and underground water with the unnamed slough, which is diverted by the Young Sump, water directly from the B.B. Harris Drain, at the point shown in No. 3, above. All water diverted by the Young Sump is tributary to Spring Creek, a tributary of the Cache La Poudre River.

5. Date of initiation of appropriation: 1894

Date water applied to beneficial use: 1894

How appropriation was initiated:

Water was applied to beneficial use for irrigation purposes. See No. 7, below.

6. Amount claimed:

0.24 c.f.s., absolute

7. Use or Proposed use:

The water diverted from the Young Sump is used to irrigate lawn, garden, and pasture on the following described property:

All land lying within the W 1/2 of the NE1/4 of the NW1/4 of Sec. 23, T 7 N, R 69W excluding the horse pasture, house, driveway and road right-of-way totalling 4.56 acres plus .06 acres in the SW1/4 of the NE1/4 of the NE1/4 of the NW 1/4 of Sec. 23, T 7 N, R 69 W.

The total number of acres irrigated is 4.62 acres.

David F. Jankowski #5787
Nossaman, Krueger & Marsh
511 16th Street, Suite 300
Denver, CO 80202
(303) 595-9441

George E. Radosevich #8622
910 15th Street, #866
Denver, CO 80202
(303) 573-5556

ATTORNEYS FOR APPLICANTS

STATE OF COLORADO )
) ss.
COUNTY OF LARIMER )

VERIFICATION

I, Robert A. Young, applicant herein and owner of the Young Sump and the property irrigated therefrom, hereby state that I have read the foregoing application, that to the best of my knowledge, information, and belief, the same is true, and I verify that it is correct.

Robert A. Young

Subscribed and sworn to before me this 28 day of December, 1981, by Robert A. Young.

My Commission expires: ______________________

Notary Public

[SEAL]

81 CW459 ROBERT A. & LYNN YOUNG, 1601 Sheely Dr., Ft. Collins, CO 80526 (David F. Jankowski, NOSSAMAN, KRUEGER & MARSH, Kittredge Bldg., 511 16th St., #300, Denver, CO 80202). Application for Water Right In Surface and Underground Water Right IN LARIMER COUNTY. Young Sump is located in the NW¼NW¼, S23, T7N, R69W, 6th P.M., 670'E of the W. line of said sec. & 515'S. of the N. line of said sec. The street address of the point of diversion is 1601 Sheely Drive, Ft. Collins, CO. Source: underground & surface water occurring naturally in an unnamed slough, water collected in the Tombough Drain & the B.B.Harris Drain. Said drains are located in S22 & 23 all in R69W, 6th P.M. Appropriation: 1894. Amount claimed: 0.24 cfs. Use: irrigate 4.62 acres lawn, garden & pasture in W½NE¼NW¼, S23. (3 pgs)

81 CW460 THE CITY OF BROOMFIELD, Attn: Harvey W. Curtis, Esq., NOSSAMAN, KRUEGER & MARSH, 511 16th Street, Suite 300, Denver, CO 80202. Application for Conditional Water Right in JEFFERSON COUNTY. Golden Ralston Creek and Church Ditch located at a point located on the North Bank of Clear Creek in the NE¼, S32, T3S, R70W, 6th P.M. - A #4 rebar and cap, L.S. 11389, set S 61°27'55" W, 1497.21' from the NE Corner of said S32; said rebar and cap is located 2.00' N and 2.00' W of the NE Corner of the Headgate of the Church Ditch. Source: Clear Creek, a tributary of the South Platte River. Appropriation: 3/25/1981. Amount claimed: 50 cfs. Proposed use of water: Use in the Broomfield municipal water supply system as it now or may hereafter exist, for all beneficial uses including municipal, domestic, irrigation, piscatorial, commercial, industrial, recreation, exchange, replacement, plans of augmentation, irrigation of city parks and open space, and other beneficial uses to which it may be placed by the City of Broomfield or those users to whom it provides water, including storage for all of the aforesaid uses in Great Western Reservoir in an amount up to 26,000 AF when the water is not needed immediately for said uses. Great Western Reservoir is located generally in the N½, S7 and the S½ of S6, T2S, R69W, 6th P.M. (2 pages; Exhibit 3 pages)

81 CW461 THE CITY OF BROOMFIELD, Attn: Harvey W. Curtis, Esq., NOSSAMAN, KRUEGER & MARSH, 511 16th Street, Suite 300, Denver, CO 80202. Application for Conditional Water Right in JEFFERSON COUNTY. McKay Ditch located on the SE Bank of Coal Creek in the SW¼, S18, T2S, R70W, 6th P.M. - A #4 rebar and cap, L.S. 11389, set N 12°18'34" E, 2706.17' from the SW Corner of said S18; said rebar and cap is located 0.20' S of the SW Corner of the Headgate of the McKay Ditch. Source: Coal Creek, a tributary of Boulder Creek a tributary of the South Platte River. Appropriation: 3/25/1981. Amount claimed: 5 cfs. Proposed use of water: Use in the Broomfield municipal water supply system as it now or may hereafter exist, for all beneficial uses including municipal, domestic, irrigation, piscatorial, commercial, industrial, recreation, exchange, replacement, plans of augmentation, irrigation of city parks and open space, and other beneficial uses to which it may be placed by the City of Broomfield or those users to whom it provides water, including storage for all of the aforesaid uses in Great Western Reservoir in an amount up to 26,000 AF when the water is not needed immediately for said uses. Great Western Reservoir is located generally in the N½, S7 and the S½ of S6, T2S, R69W, 6th P.M. (2 pages; Exhibit 3 pages)

81 CW462 THE CITY OF BROOMFIELD, Attn: Harvey W. Curtis, Esq., NOSSAMAN, KRUEGER & MARSH, 511 16th Street, Suite 300, Denver, CO 80202. Application for Conditional Water Right in BOULDER COUNTY. Community Ditch located on the SE Bank of South Boulder Creek in the SE¼ of S25, T1S, R71W, 6th P.M. - A point lying N 36°26'51" W, 1982.69' from the SE Corner of said S25. Source: South Boulder Creek, a tributary of Boulder Creek, a tributary of the South Platte River. Appropriation: 3/25/1981. Amount claimed: 25 cfs. Proposed use of water: Use in the Broomfield municipal water supply system as it now or may hereafter exist, for all beneficial uses including municipal, domestic, irrigation, piscatorial, commercial, industrial, recreation, exchange, replacement, plans of augmentation, irrigation of city parks and open space, and other beneficial uses to which it may be placed by the City of Broomfield or those users to whom it provides water, including storage, by pumping or exchange, for all of the aforesaid uses in Great Western Reservoir in an amount up to 26,000 AF when the

Case No. 81CW459 December Resume

---

SUMMARY OF CONSULTATION Held March 5, 1982

---

IN THE MATTER OF THE APPLICATION FOR WATER RIGHTS OF:

ROBERT A. & LYNN YOUNG

In Larimer County

FILED IN DISTRICT COURT 1st DIST. WATER COURT WELD COUNTY COLO. 82 MAR 30 P 1:03

---

DIVISION ENGINEER RECOMMENDATIONS

1. ____ Recommend approval, no apparent conflict.

2. X Recommend approval, with conditions (see comments)

3. ____ Recommend denial (see comments)

COMMENTS: Applicant should prove historic use. This is a very junior right.

I certify that copies hereof were mailed to the following persons at the addresses given in the pleadings

David F. Jankowski

Div + State

Date of Mailing 3-31-82

Marcie E. Spilts

~~Deputy~~ Water Clerk

James R. Clark
Division Engineer

---

The referee has reviewed these comments from the Division Engineer and will consider these recommendations but will not be bound by such recommendations.

Raymond S. Liesman
Water Referee

MAR 30 1982
Date

---

This notice is being sent to all parties pursuant to HB 1504 (1973 CRS 37-92-302(4)).

cc: Jeris Danielson
Division Office

IN THE DISTRICT COURT IN AND FOR THE
COUNTY OF WELD AND
STATE OF COLORADO

Civil Action No. 81CW459

BE IT REMEMBERED, That heretofore and on to-wit, the 1st day of March,

A. D. 1983, the same being one of the regular juridical days of the March A. D. 1983 Term of Court, the following proceedings, inter alia, were had and entered of record in said Court to-wit:

ORDER OF COURT

CONCERNING THE APPLICATION FOR WATER RIGHTS OF:

ROBERT A. & LYNN YOUNG

IN LARIMER COUNTY

Raymond S. Liesman
Referee Judge XX

Reporter

Marcie E. Spelts
Clerk

This matter coming on for Term-Day with the following determination having been made: Dave F. Jankowski to send in proof of date.

EXHIBIT F

IN THE WATER COURT IN AND FOR

WATER DIVISION NO. 1

STATE OF COLORADO

Case No. 81-CW-459

| | | |
|---|---|---|
| CONCERNING THE APPLICATION FOR WATER RIGHTS OF ROBERT A. AND LYNN YOUNG IN LARIMER COUNTY, COLORADO. | ) | AFFIDAVIT |
| TRIBUTARY INVOLVED: SPRING CREEK, TRIBUTARY TO THE CACHE LA POUDRE RIVER | ) | |

Jamia C. Riehl, being duly sworn, deposes and says:

1. I am a resident of Fort Collins, Colorado. My address in Fort Collins is P. O. Box 1207, Fort Collins, Colorado 80522.

2. On May 21, 1962, my former husband, Dr. Herbert Riehl, and I purchased the property known as 1601 Sheely Drive in Fort Collins, which is the same property now owned by Dr. and Mrs. Robert Young. We purchased the property from Inez Reeble and owned the property until 1971 when we sold it to Dr. and Mrs. Young.

3. At the time that my husband and I purchased 1601 Sheely Drive the same concrete sump on behalf of which Dr. and Mrs. Young have applied for a water right decree in this case was in existence.

4. Mrs. Reeble and her late husband had used the water of the sump to irrigate the property at 1601 Sheely Drive and, at the time we purchased said property, offered to us the opportunity to purchase the piping system which carried water from the sump for use on the property.

5. During the period in which Dr. Riehl and I lived at 1601 Sheely Drive we did use the water of the sump to irrigate that property.

6. The concrete structure constituting the sump appeared to be quite old when we purchased 1601 Sheely Drive in 1962, though I do not know the exact date of its construction.

Further Affiant saith not.

Jamia C. Riehl

Jamia C. Riehl

STATE OF COLORADO )
) ss.:
COUNTY ____________)

Subscribed and sworn to before me by Jamia C. Riehl this 18 day of December, 1982.

My commission expires: Jamia C. Riehl

My commission expires Mar. 1, 1986.

Notary Public

Address: 717 17th St. #290
Denver, Colorado 80203

IN THE WATER COURT IN AND FOR

WATER DIVISION NO. 1

STATE OF COLORADO

Case No. 81-CW-459

| | | |
|---|---|---|
| CONCERNING THE APPLICATION FOR WATER RIGHTS OF ROBERT A. AND LYNN YOUNG IN LARIMER COUNTY, COLORADO. | ) | AFFIDAVIT |
| TRIBUTARY INVOLVED: SPRING CREEK, TRIBUTARY TO THE CACHE LA POUDRE RIVER | ) | |

Paul Waag, being duly sworn, deposes and says:

1. I am a resident of Fort Collins, Colorado.

2. In 1931, I moved with my family to live on and farm all but 10 acres of the NW/4 of Section 23, Township 7 North, Range 69 West of the 6th Principal Meridian, which parcel of land is located within the City of Fort Collins, Colorado. The North boundary of said parcel is formed by Prospect Street and the West boundary by Shields Street. The 10 acres not included in this parcel were those 10 acres in the very Northwest corner of the quarter section. In 1931, said 150 acre parcel was owned by the Poudre Valley Bank, now known as United Bank, in Fort Collins, and I leased said parcel from Poudre Valley Bank. Prior to the acquisition of the parcel by the Poudre Valley Bank, the land had been owned by Mr. B. B. Harris.

3. I lived on this parcel of land with my family from 1931 to 1942 and farmed the land each of the eleven years I lived there. In 1942, we moved to our present place of residence in Fort Collins.

4. In the mid-1940s, I purchased the above-described parcel of land at auction from a Mr. Kling and continued to own it until 1950 when I sold the parcel to Mr. and Mrs. Carl Birkey.

5. When I moved to the 150 acre parcel in 1931, there existed certain underground drains on the parcel running to it from the west. One of these drains ran to my parcel in a west to east direction across the 10 acre parcel in the northwest corner of the section to my parcel. The other drain ran in a west to east direction, crossing Shields Street to my parcel. This drain ran south of and did not cross the 10 acre parcel.

6. These two drains ultimately joined at a point on my parcel north of Spring Creek.

7. From the time I began farming the 150-acre parcel in 1931, I used water from these drains to irrigate both on the north and south side of Spring Creek. I irrigated 5 to 7 acres of land north of Spring Creek and more acreage south of the Creek.

8. I obtained water from the drain system by means of small check dams within sumps approximately 5 to 6 feet deep in the drain. The check dams caused the sumps to overflow and flood the surrounding land, enabling me to irrigate.

9. I used the drain water for irrigation purposes from 1931 to 1942 when I farmed the above-described parcel and later, throughout the latter part of the forties, when I owned that parcel.

10. I did not construct the drain system described herein. Said system existed and was used to provide irrigation water prior to the time I first farmed this land in 1931.

Further Affiant saith not.

Paul Waag

Paul Waag

STATE OF COLORADO )<br>
) ss.:<br>
COUNTY Larimer )

Subscribed and sworn to before me by Paul Waag this 16th day of May, 1983.

My commission expires: July 1, 1983.

William H. Allen

Notary Public<br>
Address: 125 S. Howes<br>
Fort Collins, CO 80521

**IN THE WATER COURT**
**DIVISION I**
**STATE OF COLORADO**

July 13th, 1983

**TO:** David Jankowski, Esq.
511 16th Street, #300
Denver, Co 80202

St/Div

The Water Court Referee for Division I has instructed me to forward you this copy of his Ruling in Case No. 81CW459.

Please check your Ruling very CAREFULLY! If any errors are found, notify the Water Court IMMEDIATELY.

You have within twenty (20) days after the above date of mailing to file with the Water Clerk any Protest to the Referee's Ruling. Any protest to the Referee's Ruling must be filed on or before August 2nd, plus any additional time allowed by Rule 6(e), CRCP. In the absence of any Protest being received, the Judge of the Water Court will incorporate the Referee's Ruling into the Decree which will be entered after August 12th, 1983.

Marcie E. Spelts
Marcie E. Spelts
Water Clerk, Division No. I
POB "C"
Greeley, CO 80632

DISTRICT COURT, WATER DIVISION I, COLORADO

FILED IN DISTRICT COURT

Case No. 81 CW 459

---

FINDINGS AND RULING OF THE REFEREE AND DECREE OF THE WATER COURT

---

WELD COUNTY COLO.

CONCERNING THE APPLICATION FOR WATER RIGHTS OF:

ROBERT A. AND LYNN YOUNG in LARIMER COUNTY

---

THIS CLAIM, having been filed with the Water Clerk, Water Division I, on December 31, 1981 and the Referee being fully advised in the premises, does hereby find:

All notices required by law of the filing of this application have been fulfilled, and the Referee has jurisdiction of this application.

No statement of opposition to said application has been filed, and the time for filing such statement has expired.

All matters contained in the application having been reviewed, and testimony having been taken where such testimony is necessary, and such corrections made as are indicated by the evidence presented herein, IT IS HEREBY THE RULING OF THE WATER REFEREE:

1. The name and address of the claimant:

   Robert A. and Lynn Young
   1601 Sheely Drive
   Fort Collins, Colorado 80526

2. The name of the structure:

   Young Sump

3. The legal description of the structure:

   NW$\frac{1}{4}$NW$\frac{1}{4}$, Section 23, Township 7 North, Range 69 West of the 6th P.M., Larimer County, at a point approximately 515 feet South of the North line and 670 feet East of the West line, Section 23, a/k/a 1601 Sheely Drive, Fort Collins.

4. The source of water:

   Natural runoff occurring in an unnamed slough and underground water collected in the Tombough Drain and the B. B. Harris Drain. Said drains are located in sections 22 and 23.

   All tributary to Spring Creek, a tributary of the Cache La Poudre River.

5. The date of appropriation:

   June 30, 1930

Case No. 81 CW 459
Page 2
Young

6. The amount of water:

0.24 cubic feet per second

7. The use of the water:

Irrigation on 4.62 acres in W½NE¼NW¼, Section 23.

8. The priority herein awarded said Young Sump was filed in the Water Court in the year of 1981 and shall be administered as having been filed in that year; and shall be junior to all priorities filed in previous years. As between all rights, filed in the same calendar year, priorities shall be determined by historical dates of appropriation and not affected by the date of entry of ruling.

DATED this 13th day of July, 1983.

Raymond S. Liesman
RAYMOND S. LIESMAN
Water Referee
Water Division No. I

THE COURT FINDS: NO PROTEST WAS FILED IN THIS MATTER.

THE FOREGOING RULING IS CONFIRMED AND APPROVED, AND IS HEREBY MADE THE JUDGMENT AND DECREE OF THIS COURT.

Dated: ______________________

______________________
ROBERT A. BEHRMAN
Water Judge
Water Division No. I
State of Colorado

DISTRICT COURT
WATER DIVISION NO. 1
STATE OF COLORADO
P.O. BOX "C"
GREELEY, COLORADO 80632
356-4000

ROBERT A. BEHRMAN, JUDGE

RAYMOND S. LIESMAN, REFEREE
MARCIE E. SPELTS, CLERK

TO WHOM IT MAY CONCERN:

The attached has been enclosed for the following reason:

( ) Ruling was revised.

( ) Decree was amended.

( ) Proposed Decree has been signed by the Judge, and what is attached is a copy of the signature page.

( ) Replacement page(s) we ask to be substituted

( ) Typographical error has been corrected pursuant to correspondence received in this office.

(xx) Ruling has been decreed and what is attached is a copy of the signature page showing the Judge's signature to replace the last page of the ruling you had previously received.

Should you have any questions, please do not hesitate contacting this office.

Very truly yours,

Marcie E. Spelts

(Mrs.) Marcie E. Spelts
Water Clerk
Water Division No. I

Case No. 81 CW 459
Page 2
Young

6. The amount of water:

0.24 cubic feet per second

7. The use of the water:

Irrigation on 4.62 acres in NE¼NW¼, Section 23.

8. The priority herein awarded said Young Sump was filed in the Water Court in the year of 1981 and shall be administered as having been filed in that year; and shall be junior to all priorities filed in previous years. As between all rights, filed in the same calendar year, priorities shall be determined by historical dates of appropriation and not affected by the date of entry of ruling.

DATED this 13th day of July, 1983.

RAYMOND S. LIESMAN
Water Referee
Water Division No. I

THE COURT FINDS: NO PROTEST WAS FILED IN THIS MATTER.

THE FOREGOING RULING IS CONFIRMED AND APPROVED, AND IS HEREBY MADE THE JUDGMENT AND DECREE OF THIS COURT.

Dated: AUG 2 9 1983

ROBERT A. BEHRMAN
Water Judge
Water Division No. I
State of Colorado

## Administrative Action—The Wyoming Example

Perfection of water rights in Wyoming requires administrative action. For rights to surface stream water, a permit issuance followed by award of a certificate is required. Perfection of rights to use other types of water follows a similar, but not identical, process. In Wyoming, as in most western states, the Agricultural Extension Service publishes brief summaries of the applicable law. In Wyoming, it is entitled *Brief Summary of Wyoming Water Law,* Bulletin 531, October 1970, Agricultural Extension Service, University of Wyoming, Laramie, and *Everybody's Guide to Wyoming Water Administration,* Bulletin 530.

Based on major watershed boundaries, the state is divided into four water divisions, each having its own division superintendent (working under the state engineer) for the perfection and administration of water rights. The perfection or permit/certification process now existing in Wyoming was adopted with the

State Constitution in 1890, based on the pioneering work of Elwood Mead. This system, with a few minor revisions, has worked well and was the basis for similar systems in many western states.

The Wyoming perfection process for rights to use surface stream water may be outlined as follows:

1. Anyone wanting to use water must first apply for a permit from the state engineer on forms available from his office in Cheyenne.
2. An engineer or surveyor, qualified to practice under Wyoming law, must make a survey and prepare the maps and plans necessary for the application.
3. The application form, maps and plans, and a two dollar filing fee are submitted to the state engineer. The priority date is established by the date of receipt in the state engineer's office.
4. Upon approval of the application, the state engineer issues a permit authorizing development of the proposed water project.
5. Construction of the project must be started within one year after the permit is issued.
6. Project construction must be completed and water put to beneficial use within the times specified on the approved permit.
7. The permittee must formally notify the state engineer, in writing, of the dates construction began, when construction was completed, and when water was put to beneficial use.
8. If the permittee cannot begin, complete, and/or put the water to use in the time prescribed, he may ask the state engineer to extend any or all of the time limits. He must cite good cause for an extension and request it before expiration of the original time limits. If a time extension is granted, the date of priority remains the same.
9. After the beneficial use of water is made or a reservoir constructed and the required notices submitted, a final proof of appropriation or construction is submitted to the Board of Control (consisting of the state engineer and the four division superintendents).
10. This proof is advertised in the local newspaper and an inspection of the project made by the division superintendent.
11. If everything is in order and no protests are filed, a Certificate of Appropriation or of Construction is issued by the Board and recorded in the county clerk's office where the project is located. This is evidence of an adjudicated water right.
12. If the permittee is dissatisfied with the actions of the Board of Control, he may appeal their decision to the courts.

Following this discussion, we reproduce the salient documents in the perfection of a Wyoming water right, the Good Ditch No. 1, diverting from a tributary of the Cheyenne River in Niobrara County, Wyoming. In reviewing these materials from the state engineer's files, note the following attributes of each document:

1. The application was filed on April 4, 1972 with the state engineer's office. This date will be the priority date of the water right if the perfection process continues without interruption. The map submitted with the application was prepared by J. H. Coffman, an engineer and surveyor. In contrast to the Colorado application described in the previous section, this application was submitted by the engineer—a lawyer was not involved. On April 22, less than three weeks after the application was filed, it was approved by the state engineer, subject to the requirements that construction begin within a year and be completed by December 31, 1974, together with the diversion and application of water to beneficial use. The last page of the material accompanying the application, prior to the map, shows the permit status following its filing and approval.
2. The state engineer's permit approval required that construction on the ditch must begin within one year, or by April 22, 1973. On December 29, 1972, the state engineer sent the applicant, Mr. Slagle, a friendly reminder that a Notice of the Commencement of Work had not been received, and should it not be received by April 22, 1973, the permit would expire. The state engineer provided a form for that notice (already partially filled in) and indicated that an extension of time could be obtained.
3. On April 16, 1973, the state engineer received a Notice of Commencement of Work indicating that work had commenced on March 2, 1973, a month and one-half before the permit would otherwise have expired.
4. The state engineer's permit approval required that, by December 31, 1974, work on the ditch be completed and water diverted and put to beneficial use. On September 30, 1974, the state engineer again wrote to Mr. Slagle indicating that Notice of Completion of Construction and the Application of Water to Beneficial Use had not been received. He again provided forms for both notices and indicated that an extension could be obtained.
5. On December 12, 1974, the state engineer received two documents: a Notice of Completion of Construction for the ditch and a letter requesting an extension of time to apply water to beneficial use, since water would not be available to divert until spring. At the bottom of the letter is a hand-written note by JHA (assistant state engineer) that time for filing the Notice of Beneficial Use had been extended for one year, until December 31, 1975.
6. On September 30, 1975, the state engineer again wrote Mr. Slagle indicating that he had not yet received Notice of Application of the Water through the Ditch to Beneficial Use. He again provided the form to make such notice and indicated that an extension of time might be possible.
7. On December 31, 1975, the last possible date absent an extension, the state engineer received a Notice of Beneficial Use.

8. The original permit required that proof of appropriation be submitted to the Board of Control within five years after the water was applied to beneficial use, which was on June 8, 1975. Although the proof was not due until 1980, Mr. Slagle made his proof in March 1978. On the proof form, paragraphs 1 through 9 are completed by the permittee, while the remainder of the form, including an attached inspection report, is completed by the division superintendent, who recommended approval in this case.
9. A Certificate of Appropriation issued by the Board of Control in May of 1980 is also found in this section. There are two interesting things about this certificate: the date of appropriation (and priority) is April 4, 1972, the date that the permit application was received by the state engineer and although the permit application indicated that the carrying capacity of the ditch was 63 cfs, only 1.34 cfs was the subject of the certificate. This is because the duty of water in Wyoming, as stated above the state engineer's signature on the permit approval, is limited to 1 cfs for each 70 acres of land irrigated. Dividing the total number of acres involved, 94.4, by 70 ac/cfs, results in 1.34 cfs. Regardless of the amount of water actually used or needed for the irrigation of this land, the amount of its water right is limited by statute to 1.34 cfs.

The Wyoming system for perfecting water rights has been called unsophisticated and cumbersome. The forms are not masterpieces of graphic design and the paperwork could be tidier; nevertheless, the system does work and that is what is important. The state engineer's office generally maintains a cordial relationship with water users. Other states which have adopted the basics of the Wyoming system may have fancier forms and more professional-looking records, but few operate more effectively than Wyoming. The personal and friendly touch maintained to this day under the original Wyoming system is a breath of fresh air.

MICRO FILMED MAY 1 '72

NOTE: "DO NOT FOLD THIS FORM. –ONLY FORMS COMPLETED WITH TYPEWRITER OR NEATLY LETTERED WITH BLACK WATER-PROOF INK WILL BE ACCEPTED."

Form A-1
Rev. 3-7'

THIS SECTION IS NOT TO BE FILLED IN BY APPLICANT

Filing/Priority Date

THE STATE OF WYOMING }
STATE ENGINEER'S OFFICE } SS.

This instrument was received and filed for record on the 4 day of April, A.D. 19 72, at 9:00 o'clock A. M.

James H. Aduddell, Asst. State Engineer

Recorded in Book 8E of Applications, on Page 89. Fee Paid $ 2.00. Map Filed D.

# STATE OF WYOMING

## OFFICE OF THE STATE ENGINEER

APPLICATION FOR PERMIT TO APPROPRIATE SURFACE WATER

WATER DIVISION NO. 2

PERMIT NO. 23810
D-13

DISTRICT NO. 1

Temp. Filing No. 21 2/99

NAME OF FACILITY GOOD DITCH NO. 1

I, J. H. Coffman of Torrington (City or Town), County of Goshen State of Wyoming, for myself, or in behalf of the applicant or applicants named in Item 1, do say:

1. The name(s) and complete mailing address(es) of the applicant(s) is/are Earl A. Slagle, Jr. ~~Lance Creek, Wyoming 82222~~ Rt. 2, Newcastle, Wyo. 82701
(If more than one applicant, designate one to act as Agent for the others)
2. The use to which the water is to be applied is irrigation
3. The source of the proposed appropriation is Corral Draw, a tributary of West Prong of Bull Creek, a tributary of Bull Creek, a tributary of Cow Creek, a tributary of Lance Creek, a tributary of the South Fork of the Cheyenne River which is a tributary of the Cheyenne River
4. The point of diversion of the proposed works is located No.40°56'E. 1389 feet distant from the SW corner of Section 13 T. 39 N., R. 67 W., and is in the SW¼SW¼ of Section 13 T. 39 N., R. 67 W.
5. The said ditch, canal, pipeline or other facility is to be 1.5 miles long.
6. (a) The carrying capacity of the ditch, canal, pipeline or other facility at the point of diversion is 63.04 cubic feet per second.
   (b) If pipeline, size of pipe - - inches.
7. The estimated cost of said work is $2,000.00 Dollars.
8. Construction work will begin within 1 year from date of approval of this application.
9. The estimated time required for completion of the works is 1 year(s).
10. The estimated time required to complete the application of water to the beneficial uses stated in this application is 1 year(s).
11. The accompanying map is prepared in accordance with the State Engineer's Manual of Regulations and Instructions for filing applications and is hereby declared a part of this application.

Permit No. 23810

Page No. 89 (Leave Blank)

**12. The land to be irrigated under this permit is described in the following tabulation: (Give irrigable acreage in each 40-acre subdivision. Designate ownership of land, federal, State or private. If private, give names of owners.) If application is for stock or domestic purposes, indicate point of use in 40-acre subdivision.**

| Township | Range | Sec. | NE¼ | | | | NW¼ | | | | SW¼ | | | | SE¼ | | | | TOTALS |
|---|---|---|---|---|---|---|---|---|---|---|---|---|---|---|---|---|---|---|---|
| | | | NE¼ | NW¼ | SW¼ | SE¼ | NE¼ | NW¼ | SW¼ | SE¼ | NE¼ | NW¼ | SW¼ | SE¼ | NE¼ | NW¼ | SW¼ | SE¼ | |
| | | | ORIGINAL SUPPLY FOR FOLLOWING LANDS: | | | | | | | | | | | | | | | | |
| | | | | | | OWNER: EARL A. SLAGLE, JR. | | | | | | | | | | | | | |
| 39 | 67 | 24 | | | | | | | | | 26.4 | 13.3 | 7.3 | 18.0 | | 4.6 | 24.8 | | 94.4 |

Number of acres to receive original supply 94.4

Number of acres to receive supplemental supply 0

Total Number of acres to be irrigated 94.4

**REMARKS**

Under penalties of perjury, I declare that I have examined this application and to the best of my knowledge and belief it is true, correct and complete.

J. H. Coffman
Signature of Applicant or Agent

April 3, 1972
Date

**THE STATE OF WYOMING,** }
**STATE ENGINEER'S OFFICE** } **SS.**

**THIS IS TO CERTIFY that I have examined the foregoing application and do hereby grant ~~reject~~ the same subject to the following limitations and conditions:**

**This permit grants only the right to use the water available in the stream after all prior rights are satisfied.**

**Construction of proposed work shall begin within 1 year from the date of approval.**

**The time for completing the work shall terminate on December 31, 19**74.

**The time for completing the application of water to beneficial use shall terminate on December 31, 19**74**, and final proof of appropriation shall be made within 5 years thereafter.**

**The amount of appropriation shall be limited to 1 cubic foot per second of time for each 70 acres of land reclaimed on or before December 31, 19**74**, except as provided in Section 41-184, Wyoming Statutes, 1957, ~~and the additional volume used for purposes on or before said date.~~**

**Witness my hand this** 22nd **day of** April**, A. D. 19**72

Floyd A. Bishop
**FLOYD A. BISHOP, State Engineer.**

**Permit No.** 23810

**Page No.** 89
**(Leave Blank)**

**PERMIT NO. 23810**
D-13

**PERMIT STATUS**

Priority Date April 4, 1972 Approval Date April 22, 1972

December 29, 1972 - Notice of expiration of time for commencement mailed. MICRO-FILMED DEC 18'72

April 16, 1973 - Notice of Commencement of Work, March 2, 1973, received. MICRO-FILMED APR 23'73

September 30, 1974 - Notice of expiration of time for completion of construction and completion of beneficial use mailed. MICRO FILMED SEP 9'74

December 12, 1974 - Notice of Completion of Construction, December 5, 1974 received.

December 12, 1974 - EXTENSION GRANTED, IN RESPONSE TO REQUEST RECEIVED December 12, 1974. Time for completion of beneficial use has been extended to December 31, 1975.

JAMES H. ADUDDELL
Assistant State Engineer '74

September 30, 1975 - Notice of expiration of time for completion of beneficial use mailed. SEP 24'75

December 31, 1975 - Notice of Completion of Beneficial Use, June 8, 1975 received JAN 14'76

CERT. REC. 73, P. 283 PROOF NO. 32527
IRR. X STK. ____ DOM. ____ MISC. ____ MICRO-FILMED NOV 26'80
AC. 94.4 C.F.S. 1.34

**NOTICE**

A Manual of Regulations and Instructions for filing applications will be furnished by the State Engineer's Office upon request. By carefully complying with the instructions contained in the Manual, much trouble and delay will be saved the applicant, the professional engineer or land surveyor, and the State Engineer's Office.

This application must be accompanied by maps in duplicate, prepared in accordance with the Manual, and by a filing fee of Two Dollars ($2.00).

Applications returned for corrections must be resubmitted to the State Engineer within 90 days with the corrections properly made; otherwise the filing will be canceled.

This application, when approved, does not constitute a complete water right. It is your authority to begin construction work, which must be commenced within one year from the date of approval of this application.

Notice of commencement of work, completion of the work, and of application of the water to the beneficial uses described in the permit, must be filed in the State Engineer's Office before the expiration of the time allowed in the permit.

If extensions of time beyond the time limits set forth in the permit are required, requests for same must be in writing, stating why the additional time is required, and must be received in the State Engineer's Office before the expiration of the time allowed in the permit.

To perfect your water right, notify your Water Division Superintendent that you are ready to submit final proof. This notice should be sent to the Superintendent as soon as possible after you apply the water to the beneficial uses described in your permit. When you have submitted your proof before the Superintendent, it will be considered by the State Board of Control, and, if found to be satisfactory, the Board will issue to you a Certificate of Appropriation which will constitute a completed water right.

The granting of a permit does not constitute the granting of right-of-way. If any right-of-way is necessary in connection with the application it should be understood that this responsibility is the applicant's.

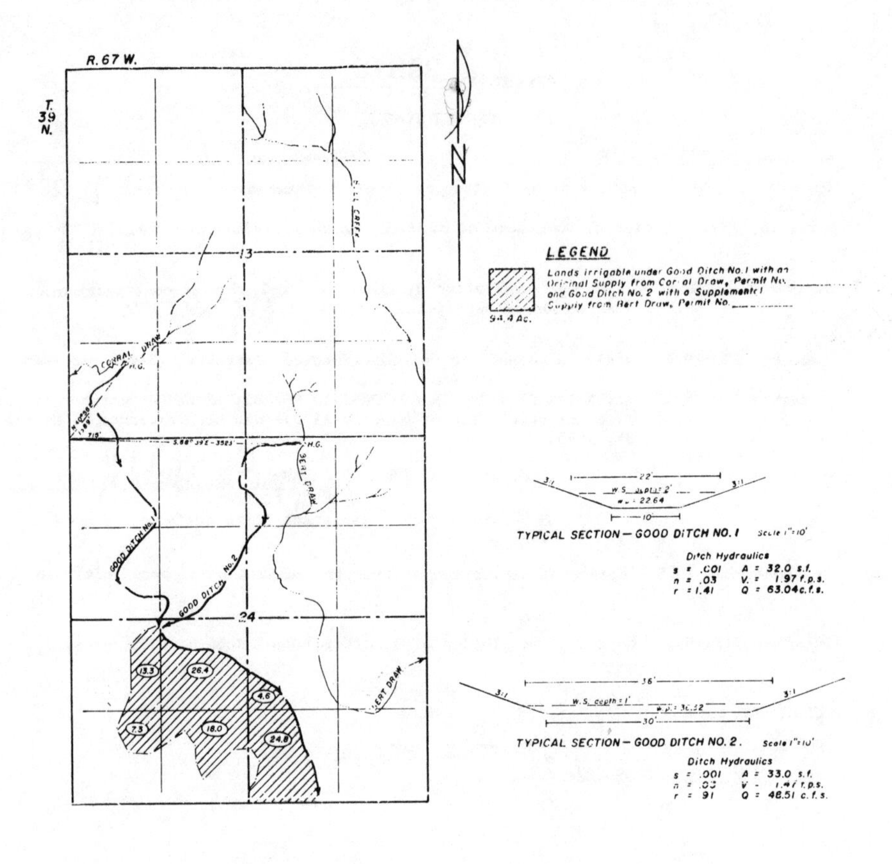


CERTIFICATE OF SURVEYOR

STATE OF WYOMING ) SS.
COUNTY OF GOSHEN (

I, J. H. COFFMAN, OF TORRINGTON, WYOMING, HEREBY CERTIFY THAT THIS MAP WAS MADE FROM NOTES TAKEN DURING AN ACTUAL SURVEY MADE BY ME ON MARCH 9, 1972, AND THAT IT CORRECTLY REPRESENTS THE PROPOSED FACILITY, THE LOCATION AND AREA OF THE LANDS PROPOSED TO BE IRRIGATED IN EACH SUBDIVISION, AND THE LOCATION OF STREAMS, AS DESCRIBED IN THE ACCOMPANYING APPLICATION.

I FURTHER CERTIFY THAT COPIES OF RECORDS ON FILE IN THE OFFICES OF THE WYOMING STATE ENGINEER, THE US-BLM., AND THE US-ASCS. WERE OBTAINED AND USED IN THE PREPARATION OF THIS MAP. US-SCS. FIELD NOTES WERE ALSO USED IN THE PREPARATION OF THIS MAP.

J. H. Coffman
ENGINEER SURVEYOR

WYOMING
REGISTRATION NO. PE-LS 529

MAP
TO ACCOMPANY APPLICATIONS
FOR
GOOD DITCH NO. 1
&
GOOD DITCH NO. 2

APPROVED April 22, 1972
Floyd A. Bishop
FLOYD A. BISHOP STATE ENGINEER

Applicant: Earl A. Slagle Jr.
Lance Creek,
Wyoming — 82222

Scale 1" = 1000'

21 2/99 And 21 3/99

THIS PRINT MAY NOT BE TRUE TO SCALE
0 1 2 3 4 5 6

THE STATE OF WYOMING

Form SE-4 (Rev. 70)

*State Engineer's Office*

STATE OFFICE BUILDING CHEYENNE, WYOMING 82001

December 29, 1972

Earl A. Slagle, Jr.
Lance Creek, Wyoming 82222

cc: J. H. Coffman, Engineer
Box 476
Torrington, Wyoming 82240

Dear Sir:

The records of this office show that the time for commencement of work on the Good Ditch No. 1 under Permit No. 23810 expires on the 22nd day of April 19 73 .

The law requires that actual construction work be commenced within one year after the date of approval of the permit, and that notice of the date when the work was started be submitted to the State Engineer.

If you need more time in which to commence work under the above mentioned permit, you must write the State Engineer, before the date of expiration shown above, and ask for an extension of time. Good reasons must be given in support of such a request. Such request for an extension of time would be for commencement of work only.

Floyd A. Bishop
State Engineer

CERTIFIED

(Complete and return form below)

NOTICE OF COMMENCEMENT OF WORK

__________, Wyo. __________, 19__

State Engineer
Cheyenne, Wyoming 82001

I hereby notify you that work was commenced on the Good Ditch No. 1 proposing to divert water from Corral Draw, trib. West Prong Bull Creek under Permit No. 23810, said work having been commenced on the _____ day of ________ 19__.

____________________
Signature of Applicant
or Authorized Agent

---

NOTICE OF COMMENCEMENT OF WORK

Lance Creek, Wyo. April 11, 1973

State Engineer
Cheyenne, Wyoming 82001

I hereby notify you that work was commenced on the Good Ditch No. 1 proposing to divert water from Corral Draw, trib. West Prong Bull Creek under Permit No. 23810, said work having been commenced on the 3rd day of March 1973.

RECEIVED APR 13 1973

MICRO-FILMED APR 23 '73

Earl A. Slagle Jr.
Signature of Applicant
or Authorized Agent

THE STATE OF WYOMING

## State Engineer's Office

STATE OFFICE BUILDING CHEYENNE, WYOMING 82001

September 30, 1974

Earl A. Slagle, Jr.
Lance Creek, Wyoming 82222

cc: J. H. Coffman, Engineer
Box 476
Torrington, Wyoming 82240

Dear Mr. Slagle:

The records of this office show that the time for the completion of construction and completion of the application of water to beneficial use through the Good Ditch No. 1, Permit No. 23810, expires on the 31st day of December, 1974.

Failure to complete the construction or make beneficial use of water before the permit expires, acts as a forfeiture of the water right granted by the permit. If the facility described in said permit has been completed and beneficial use of the water has been made within the terms of this permit, notice should be sent to this office before the expiration of the construction and beneficial use period, using the form attached for that purpose.

The State Engineer is given authority to extend permits, within certain limits, and for good cause shown. If more time is desired, such additional time should be requested in writing before the permit expires.

NOTICE OF COMPLETION OF CONSTRUCTION MEANS THAT ALL CONSTRUCTION WORK HAS BEEN COMPLETED.

NOTICE OF APPLICATION OF WATER TO BENEFICIAL USE MEANS THAT ALL INTENDED LANDS HAVE BEEN IRRIGATED OR BENEFICIAL USE MADE AT ALL THE PROPOSED POINTS OF USE.

Sincerely,

CERTIFIED

Floyd A. Bishop
FLOYD A. BISHOP, State Engineer

---

**NOTICE OF COMPLETION OF CONSTRUCTION**

_______________ Wyo., ___________ 19___

State Engineer,
Cheyenne, Wyoming

I hereby notify you that I completed the construction of Good Ditch No. 1, Permit No. 23810, proposing to divert water from Corral Draw, trib. West Prong Bull Creek ____________ on the ________ day of ______________, 19___.

_______________________
Signature of Applicant or
Authorized Agent

---

**NOTICE OF BENEFICIAL USE OF WATER**

_______________ Wyo., ___________ 19___

State Engineer,
Cheyenne, Wyoming

I hereby notify you that I completed the application of water to beneficial use for the Good Ditch No. 1, Permit No. 23810 proposing to divert water from Corral Draw, trib. West Prong Bull Creek ____________ on the ________ day of ______________, 19___.

_______________________
Signature of Applicant or
Authorized Agent

NOTICE OF COMPLETION OF CONSTRUCTION

Newcastle Wyo., Dec. 10 1974

State Engineer,
Cheyenne, Wyoming

DEC 12 1974

I hereby notify you that I completed the construction of Good Ditch No. 1 , Permit No. 23810 , proposing to divert water from Corral Draw, trib. West Prong Bull Creek on the 5th day of Dec , 1974.

Earl A. Slagle Jr.
Signature of Applicant or
Authorized Agent

---

EXTENSION GRANTED
Permit Nos. 23810,
7224 Stk. Res.,
7225 Stk. Res.,
7227 Stk. Res.,
7228 Stk. Res.

Rt. 2, Newcastle
Wyo. 82701
Dec. 10, 1974

State of Wyoming
State Engineer's Office
Cheyenne, Wyo. 82002

Dear Sir:

I am enclosing the notice of completion of Good Ditch No. 1, but would like an extention on the beneficial use of water. This ditch depends on using run off water to work and this probably will not be available until this spring.

Also would it be possible to get an extention on:

Allen No. 2 Stock Reservoir -- Permit No. 7224 Stk. Res.
Bull No. 2 Stock Reservoir -- Permit No. 7225 Stk. Res.
Harry No. 1 Stock Reservoir -- Permit No. 7227 Stk. Res.
East Calf No. 1 Stock Res. -- Permit No. 7228 Stk. Res.

I finally got the pipe for these reservoirs but due to the dry summer I am finishing some reservoirs that I am needing worse, but hope to finish these to catch the rains this spring.

Thank you very much,

Earl A. Slagle Jr.
Earl A. Slagle, Jr.

Enclosure: 1

[handwritten:] Marge
Please extend Per # 23810 for 3 U. Use one yr to Dec. 31, 1975. Also extend Permit Nos. 7224 S.R., 7225 S.R., 7227 S.R. and 7228 S.R. for Compl. to Dec. 31, 1975 [initials]

DEC 12 1974
STATE ENGINEER
Cheyenne, Wyo.

THE STATE OF WYOMING

ED HERSCHLER
GOVERNOR

# State Engineer's Office

STATE OFFICE BUILDING EAST CHEYENNE, WYOMING 82002

September 30, 1975

Earl A. Slagle, Jr.
Route 2
Newcastle, Wyoming 82701

Dear Mr. Slagle:

The records of this office show that the time for the ~~completion of construction~~ completion of the application of water to beneficial use through the Good Ditch No. 1, Permit No. 23810, expires on the 31st day of December, 19 75.

Failure to complete the construction or make beneficial use of water before the permit expires, acts as a forfeiture of the water right granted by the permit. If the facility described in said permit has been completed and beneficial use of the water has been made within the terms of this permit, notice should be sent to this office before the expiration of the construction and beneficial use period, using the form attached for that purpose.

The State Engineer is given authority to extend permits, within certain limits, and for good cause shown. If more time is desired, such additional time should be requested in writing before the permit expires.

NOTICE OF COMPLETION OF CONSTRUCTION MEANS THAT ALL CONSTRUCTION WORK HAS BEEN COMPLETED.

NOTICE OF APPLICATION OF WATER TO BENEFICIAL USE MEANS THAT ALL INTENDED LANDS HAVE BEEN IRRIGATED OR BENEFICIAL USE MADE AT ALL THE PROPOSED POINTS OF USE.

CERTIFIED

George L. Christopulos
GEORGE L. CHRISTOPULOS
State Engineer

(Complete and return form below)

~~NOTICE OF COMPLETION OF CONSTRUCTION~~

_______________, Wyo., _______________, 19___

State Engineer
Cheyenne, Wyoming 82002

I hereby notify you that I completed the construction of _______________, Permit No. _______________, proposing to divert water from _______________ on the _______ day of _______________, 19___.

_______________
Signature of Applicant or
Authorized Agent

NOTICE OF BENEFICIAL USE OF WATER

_______________, Wyo., _______________, 19___

State Engineer
Cheyenne, Wyoming 82002

Permit No. 23810

I hereby notify you that I completed the application of water to beneficial use for the Good Ditch No. 1, proposing to divert water from Corral Draw, trib. West Prong Bull Creek _______________ on the _______ day of _______________, 19___.

_______________
Signature of Applicant or
Authorized Agent

**NOTICE OF BENEFICIAL USE OF WATER** 1476

Newcastle, Wyo., Dec 29, 1975

State Engineer
Cheyenne, Wyoming 82002

Permit No. 88810.

I hereby notify you that I completed the application of water to beneficial use for the Good Ditch No. 1, proposing to divert water from Corral Draw, trib. West Prong Bull Creek on the 8th day of June, 1975.

DEC 31 1975
STATE ENGINEER
Cheyenne, Wyo.

[signature]
Signature of Applicant or Authorized Agent

MICRO-FILMED NOV 26 '80

32527

# STATE OF WYOMING

## PROOF OF APPROPRIATION OF WATER

**NOTE:** Use Typewriter or Ball Point Pen

**PERMIT NO.** 23810 **DATE OF PRIORITY** April 4, 1972

**NAME OF DITCH OR CANAL** Good Ditch No. 1

**SOURCE OF WATER SUPPLY** Corral Draw, trib. West Prong Bull Creek, trib. Bull Creek,

**TRIBUTARY OF** Cow Creek, trib. Lance Creek, trib. South Fork Cheyenne River, trib. Cheyenne River/

1. Name of claimant
2. Postoffice ____ State ____ Zip ____
3. For what purpose is water used, Irrigation
4. (a) Is a good headgate provided? N/A Give Location, Sec. 13 T. 39 N., R. 67 W. SW¼SW¼
   (b) Have terms of permit been complied with as to width, depth and grade of ditch? yes
   (c) Was this ditch or canal built in the same location as shown on the map which accompanied the permit? yes
   (d) Have all notices required in connection with this permit been filed in the State Engineer's Office yes
5. Give legal subdivisions of land owned by you on which water has been used, and if an appropriation of water for irrigation is claimed, give the acreage which has been irrigated in each legal subdivision in compliance with terms of this permit.

| Twp. | Range | Sec. | NE¼ | | | | NW¼ | | | | SW¼ | | | | SE¼ | | | | TOTAL |
|---|---|---|---|---|---|---|---|---|---|---|---|---|---|---|---|---|---|---|---|
| | | | NE¼ | NW¼ | SW¼ | SE¼ | NE¼ | NW¼ | SW¼ | SE¼ | NE¼ | NW¼ | SW¼ | SE¼ | NE¼ | NW¼ | SW¼ | SE¼ | |
| 39 | 67 | 24 | | | | | | | | | 26A | 13.3 | 7.3 | 18 | | 4.6 | 24.8 | | 94A |

6. In what county are these lands situated? Niobrara
7. Has water been used beneficially and all the land shown in the above tabulation irrigated each year since completion of works? yes If not, state exceptions and reasons therefor
8. What documentary evidence is attached showing your ownership or control of above lands? Certificate of ownership
9. THE STATE OF WYOMING } ss. County of Niobrara

   Division No. 2
   District No. 1

   I, Earl A. Slagle, being first duly sworn, do depose and say that I have read the above proof of appropriation of water; that I understand the contents thereof; and the facts stated herein are true.

   Subscribed and sworn to before me this 30 day of March, 1978

   x Earl A. Slagle Jr.

   Paul J. Kawulok
   Superintendent, Water Division No. 2

10. Report of Division Superintendent. Field Inspection on 30 day of March, 1978
    Description of Headgate NA
    Description of Ditch or Canal Earthen
    Is this a pump diversion? no If yes, what is brand and model no. of pump? —
    Type of power — Horsepower — Size of intake line —
    Size of discharge line —
    Describe how water is used Flood irrigation
    Were all conditions of permit fulfilled? yes If not, describe deficiencies
    Do you recommend approval?
    Other Comments

**Note: Also check for attached, detailed inspection report.**

**Filed in the office of the State Board of Control**

7th day of June, 1978

William Long
Ex-officio Secretary, State Board of Control

Fees Paid $2.00 Rec. No. 7629

**Fees Paid $** 2.00

30 day of March, 1978

Paul J. Kawulok
Superintendent Water Division No. 2

Orig.—#1—Board of Control

THE MILLS COMPANY, SHERIDAN 131681

MICRO-FILMED NOV 26 '80

32527

PROOF OF APPROPRIATION INSPECTION REPORT

PERMIT NO. 23810 OWNER Earl A. Slagle Jr.

NAME OF DITCH OR CANAL Good Ditch No. 1

1. If domestic, is this the only source? NA Is water actually in use at the dwelling? NA More than one dwelling? NA For irrigating lawns, trees and garden? NA If stock use is included, are there facilities for this? ______ Describe NA

2. Describe Head Gate NA
   Measuring Device NA

3. If pump diversion - Make and Model NA Type NA
   Size Intake NA in. Size Discharge NA in. Length of Pipeline NA Type of Power NA H. P. NA

4. a. Dimensions of Ditch, Width at top 22 ft. Width at bottom 10 ft.
   Depth 2 ft. Where Measured Above irrigated Ground
   b. Size of pipe, if pipeline NA

5. If there are any changes from the terms of permit, explain NO

6. Has entire acreage been irrigated? Yes If not, estimate acreage not irrigated and mark on the map.
   a. Has or will applicant sign an elimination form for non-irrigated acres? NA
   b. What use is made of the land? Pasture

7. How is water applied to the land? Spreader Flooding

8. Have you explained to the applicant that a certificate of ownership will be necessary? Yes

9. Do you recommend approval? Yes ✓ If not, why ______

COMMENTS ______

Date of inspection 30 Mar. 1978 Signed Paul J. Kawulok - Supt.

Title Water Div. No. 2

# THE STATE OF WYOMING
## Certificate of Appropriation of Water

Proof No. 32527
Certificate Record No. 73, Page 283
Water Division No. 2, District No. 1

WHEREAS, Earl A. Slagle, Jr. and Betty A. Slagle, husband and wife have presented to the Board of Control of the State of Wyoming proof of the appropriation of water from Corral Draw, tributary West Prong Bull Creek, tributary Bull Creek, tributary Cow Creek, tributary Lance Creek, tributary South Fork Cheyenne River, tributary Cheyenne River through the Good No. 1 Ditch under Permit No. 23810 for irrigation of the lands herein described, lying and being in Niobrara County, Wyoming.

NOW KNOW YE, That the State Board of Control, under the provisions of the Statutes of Wyoming, has, by an order duly made 16th May, 1980 and entered on the 23rd day of May, A. D. 1980, in Order Record No. 23, Page 518, determined and established the priority and amount of such appropriation as follows:

Name of Appropriator Earl A. Slagle, Jr. and Betty A. Slagle, husband and wife; Postoffice Address Route 2, Newcastle 82701, Wyoming;

Date of Appropriation April 4, 1972; Total Acreage Ninety-four and four tenths (94.4) acres;

Amount of Appropriation 1.34 cu. ft. per sec.; Description of land to be irrigated and for which this appropriation is determined and established:

| TWP. | RANGE | SEC. | NE¼ | | | | NW¼ | | | | SW¼ | | | | SE¼ | | | | TOTAL |
|---|---|---|---|---|---|---|---|---|---|---|---|---|---|---|---|---|---|---|---|
| | | | NE¼ | NW¼ | SW¼ | SE¼ | NE¼ | NW¼ | SW¼ | SE¼ | NE¼ | NW¼ | SW¼ | SE¼ | NE¼ | NW¼ | SW¼ | SE¼ | |
| 39N. | 67W. | 24 | | | | | | | | | 26.4 | 13.3 | 7.3 | 18 | | 4.6 | 24.8 | | 94.4 |

Head Gate: SW¼SW¼ Sec. 13-39-67

The right to water hereby confirmed and established is limited to irrigation and the use is restricted to the place where acquired and to the purpose for which acquired; rights for irrigation not to exceed one cubic foot of water per second for each seventy acres of land for which the appropriation is herein determined and established, except when there is surplus water in the stream as provided by Sections 41-181 to 41-189, Wyoming Statutes, 1957.

IN TESTIMONY WHEREOF, I, George L. Christopulos, President of the State Board of Control, have hereunto set my hand this 23rd day of May, A. D. 1980, and caused the seal of said Board to be hereunto affixed.

Attest: William Long Ex-officio Secretary.

George L. Christopulos President.

# ACQUISITION AND DISPOSITION OF A WATER RIGHT

Acquiring an existing water right is governed by detailed considerations of state law, but insofar as general observations are possible and appropriate, they are contained in this section. One should think of the acquisition or the disposition of a water right in essentially the same framework as the acquisition and disposition of any other real property, including land. Consequently, it is entirely appropriate to convey water rights, lease them, obtain an option on them, and so on.

## Severability

Perhaps the most important thing to bear in mind when dealing with the conveyancing of water rights is that, in many states water rights are severable from the lands which they were originally appropriated or perfected to serve. In some states, however, water rights are considered to be permanent appurtenances to the land and to be conveyed with the land as a matter of law. Acquisition and disposition problems arise most frequently in those states where water rights are severable.

## Title to Water Rights

In states where water rights are not severable from the land they serve, title questions usually pose no special significance in water right transactions. Title problems are endemic, however, where the water rights can be conveyed separately from the land.

Paper title or ownership becomes the initial issue in every acquisition of a water right or interest in a water right. Lawyers are inclined to describe paper title as either "marketable" or "good." Marketable, also called "merchantable," title can be established based on a review of the public real property records. Good title may not be ascertainable from the real property records but is probably safe from challenge by third persons.

In determining title to a water right, lawyers develop what is commonly referred to as a "chain of title." Based on documents of record, lawyers determine whether it is possible to trace the title to the water right from its inception to the person who now wishes to sell, lease, or convey an option in the water right. Determining the inception of a water right in a permit state is normally a simple process, since the permit itself represents the water right's inception and it is usually easy to determine the owner of the water right for which the permit was issued.

In a mandate state, such as Colorado, it is not as easy to determine who was the first owner of the water right. Conceptually, the first owner of the water right was the person who first made the appropriation, that is, took the water and applied it to beneficial use. Unfortunately, such an appropriation is a nonrecorded act for which it is highly improbable that there will be any recorded evidence. It is sometimes possible to determine the appropriator if one can find the transcripts of testimony given during the adjudication of that water right. The person who sought adjudication may have testified not only when the appropriation was made but also the identity of the appropriator. With that information it is possible to determine the first owner of the water right. In addition, under some statutory schemes, appropriators sometimes filed maps and statements with the state engineer.

Once the identity of the first owner is established, the lawyer reviewing the chain of title must locate a deed or series of deeds by which the water right in question is conveyed from the original owner to the person who purports to be the present owner. Although water rights are one of the most important aspects of the development of the western United States, conveyancing of water rights leaves much to be desired. Typically, deeds will simply convey land "together with appurtenant water rights." In that event, it must be determined whether it was the intent of the grantor to convey the water right along with the land. If the water right was used on the land before the conveyance and continued to be used on the land after the conveyance and the water right is in fact necessary for the continued use of the land, then it may be assumed that the owner of the water right intended to convey it by inclusion in the appurtenancy language. Few senior water rights in the West have a complete chain of title. It is often a business decision whether a costly quiet title action may be necessary to determine the owner of that water right.

## Historic Use

While it is essential to know that a water right which is being acquired has good or marketable *paper* title based on real property records, once paper title is determined a more practical question arises: Is the water right only a paper water right or will it entitle the new owner to actually divert or store water? Lawyers usually call on engineers for their expert assistance and opinions in determining the practical vitality of the water right. The engineer makes a historical study to determine the extent to which the water right has actually been exercised and satisfied in the past. Investigation of important water rights may extend back for many decades. Investigation of less valuable water rights is usually limited to 10 to 15 years. The engineer tries to discover two things: the historical diversions by the water right and the historical consumptive use of the water right. Since the water right's true value is determined by its historic use, these engineering determinations are extremely important. In sophisticated water right acquisitions, the eventual purchase price is tied to the amount of historic use attributable to the water right.

For example, a water right may have been perfected in the amount of 5 cfs for irrigation purposes. After a lawyer determined that the right had good or marketable paper title, an engineering investigation might disclose that the water right had never, during the period of the engineering study, diverted more than

3 cfs. In spite of the perfected amount of 5 cfs, the practical amount associated with the water right is its maximum historical use of 3 cfs. In addition, if the engineering study determines that the water right actually diverted 3 cfs during only an average of 30 days per year, then the value of the water right is limited by its historic annual diversions of 90 cfsd, or approximately 180 af of diversions. Further, if the engineering study should establish that only 20% of the water diverted was consumed or otherwise lost to the stream (the other 80% returning to the stream after use), the historical annual consumptive use of the water right will be 36 af. When water rights are valued at $2000 to $10,000 per consumptive acre-foot of annual yield, the water right in our example would have a total value of $72,000 to $360,000.

When a water right is represented as a 5 cfs water right, the experienced engineer and lawyer, representing the buyer, will refuse to attach any value to that water right until they have completed not only the legal investigation of paper title but the more practical investigation of historic use upon which the viability or vitality of the water right is ultimately dependent.

## Purchase and Sale

Water rights are purchased and sold in various ways. The most common arrangement is for the parties to enter into a written contract which provides the purchaser an opportunity to examine the title and the viability of the water right and upon closing for the purchaser to pay a previously agreed upon amount of money in return for the receipt of a deed to the water right. Such straightforward transactions, are usually limited to water rights of lesser value or water rights so well known that the viability is simply not in question, that is, where the purchaser need not assume a substantial risk regarding viability.

Frequently the transaction takes a more complicated form. The amount of money actually paid for the water right at closing will depend on the amount of historic consumptive use associated with the water right. When the stakes are high and the water right is being purchased for a new use or a use at a different location, the purchase price is usually conditioned or calculated on (1) the successful transfer of the water right by court or administrative proceeding to the new use or place of use as well as (2) the amount of historic consumptive use which is recognized in the transfer by the appropriate administrative agency or court. With complicated transfer proceedings consuming several years, it is not unusual for the conveyance of a water right to take three to five years from the date the contract is signed to the date of closing, when money and deed are finally exchanged.

A word or two is necessary about the deeds which are typically used for water rights conveyancing. No seller is eager to make many, if any, warranties about the quality of the title or the practical viability of a water right. Consequently, the seller will not be willing, absent substantial persuasion (usually measured in dollars), to give anything other than a quitclaim deed to the water right.

The buyer, on the other hand, wants as many assurances of title as he can get from the seller and will usually seek to obtain a general warranty deed to the water rights. Where the parties are of roughly equal bargaining strength, it is not unusual for a special warranty deed or a bargin and sale deed to be used to convey the water rights.

## Leases

Many western water right owners have come to realize the value of their property. Water rights have steadily appreciated over the years and make excellent long-term investments. Most water right owners would prefer to hold on to their water rights rather than convey them for money which will decrease in value over the years. For this reason, it is not surprising to discover a ranch or a farm which has been conveyed without all of its water rights. Canny ranchers or farmers often keep some of the best water rights for themselves, leaving the high-stepping, suede-shoe operator from the East to develop formerly irrigated land as best he can. Having reserved the water rights, the rancher or farmer who has retired has a difficult problem. The water rights must continue to be used or they may suffer the fate of abandonment or forfeiture. To provide for their legitimate use and hold on to the water rights while they continue to appreciate in value, the retired rancher usually tries to lease them—often to the new owners of the farm or ranch, at least until they are able to find water on a permanent basis.

The lease of a water right is similar to the lease of land. The term of the lease is negotiated by the parties and payments under the lease are either fixed or adjusted based on the actual performance of the water right—much as in the case of the complicated sale previously described. The one provision universally insisted upon by the lessor requires that the lessee continue to use the water rights to the maximum extent possible. If that provision does not exist or if the lessee does not exercise the water rights, the lessor faces the danger of forfeiture or abandonment. Another typical provision allows cancellation of the lease by the lessor in the event that the lessee does not continue to make the maximum possible use of the water to which he is entitled under that particular water right.

## Financing

One of life's mysteries is the general good fortune (blind luck) of bankers. Nowhere is this more apparent than in the water rights area. Most bankers aren't even aware that water rights exist. Nevertheless, millions of dollars are lent annually and secured by lands or other properties whose value is utterly dependent on paper title to and the viability of water rights. The vast majority of mortgages or deeds of trust on farm and ranch properties in the West make no mention of specific water rights. A few will use the appurtenancy language previously described and discussed. Still fewer will actually set forth the water rights as security for the purchase money mortgage. Until recently, only a few banks have been stung by this gross inadvertance and neglect. With the depressed economic conditions in the agricultural industry in the past decade, lending institutions are beginning to learn their lessons the hard way. Consequently, we can expect more thoughtful drafting of the documents for securing farm and ranch loans.

Once burned, bankers become more aggressive, especially in a state where water rights are severable from the land. A practice almost unheard of a few years ago has now become commonplace. Some bankers will actually lend money on water rights alone. This has been a shot in the arm for water rights speculators who find a water right which can be purchased for a song, sign a contract for that purchase, and borrow money from the bank to make the purchase. The same day that the speculator's purchase is closed, another sale will take place—between the speculator and a pigeon willing to pay substantially more than the price paid by the speculator. The banker has lent money for only a few hours and yet has enjoyed a substantial return on that money. The speculator has spent virtually no money at all, except the cost to borrow the purchase price for a few hours one day. The pigeon, usually a municipality, has spent a lot of money but all the costs can be passed on to the taxpayers, so that doesn't matter at all. . . .

### Condemnation

Condemnation, the forced sale of a water right at a price determined by a court, is often threatened but rarely used in the West. Condemnation is a power enjoyed by virtually all levels of government and, in some states, individuals (either to acquire water for a preferred use or to condemn a right-of-way for a water transmission facility). There have been municipal condemnations of water rights—usually accompanied by great furor. Some states have imposed substantial limitations and requirements on municipal condemnation. These limitations are often so stringent they reduce the likelihood of a successful condemnation action. The major value of condemnation power is the remarkable amount of negotiating power governmental entities attain in purchasing water rights.

## ADAPTATION OF WATER RIGHTS

As exploitation of the available water resources becomes complete, we generally define a particular water source as "overappropriated." When this occurs, it is no longer possible to enjoy a reliable supply of water based on a new appropriation. Consequently, when business economics justify it, new water requirements are satisfied by old water rights. In the western United States, this usually means that an irrigation water right is acquired and converted to a new purpose, typically municipal or industrial. As a result, the majority of current legal and engineering work in the water area involves the adaptation of irrigation water rights to serve today's needs.

## General Requirement—No Injury

The major issue in adapting old water rights to meet new needs is injury. No conversion can be made which will adversely affect other water rights in any respect, including the quantity, timing, or quality of water available to them.

## Changes

A change of water right includes

> a change in the type, place, or time of use, a change in the point of diversion, a change from a fixed point of diversion to alternate or supplemental points of diversion, a change from alternate or supplemental points of diversion to a fixed point of diversion, a change in the means of diversion, a change in the place of storage, a change from direct application to storage and subsequent application, a change from storage and subsequent application to direct application, a change from a fixed place of storage to alternate places of storage, a change from alternate places of storage to a fixed place of storage, or any combination of such changes.

Changes of water rights (as opposed to augmentation plans) *usually* involve a change where timing is not an element of injury.

Making a change is an exercise in balancing depletions. The prohibition against injury is merely a recognition that junior water right owners are entitled to the maintenance of those stream conditions which existed at the time their appropriations were made. A junior priority holder cannot be said to be injured if the change of a senior priority imposes no greater or different burden on the stream than existed before the change. Consequently, the engineering and legal issue in any change proceeding is to ensure that the new use will not change the burden on the stream, meaning the quality, quantity, or timing of water available to junior priorities. While streams may be overappropriated during irrigation season, there may be unappropriated water still available during the nonirrigation season. In such an instance, it may be possible to change an irrigation water right to a water right which operates year round without causing injury to juniors.

## Plans for Augmentation

When a stream system (including tributary groundwater) is overappropriated, not only during the irrigation season but during all or parts of the remainder of the year, it may be necessary to have a plan for augmentation if one wishes to change an irrigation water right (or any other water right with a limited period of historic use) to a year-round use. Off-season injury is usually prevented by a plan for augmentation which provides for releases from storage to augment the flow of the stream to prevent injury.

## Common Derivatives

The definition of a change of water right or a plan for augmentation includes three other commonly used devices which have traditionally been given separate names: water right exchanges, substitute supplies, and temporary loans—all of which may be formal or informal, voluntary or involuntary.

When there is no injury to the rights of others, owners of priorities in reservoirs and ditches may exchange water. The upstream reservoir may release water from storage for the benefit of a downstream ditch, the reservoir thereafter storing water under the priority of the downstream ditches. Measuring devices are usually required by statute to account for the waters exchanged.

Substitution may be carried out between any combination of ditches and reservoirs and may be either voluntary or involuntary. As long as a senior priority receives an adequate supply of water, whatever the source, it may not complain of out of priority diversions under a junior priority. For example, an upstream junior ditch may divert water out of priority if it can provide substituted water to downstream juniors. To do so, the junior might contract with a downstream reservoir to make releases to the senior headgate.

In times of need, neighbors often temporarily loan their water right. As long as no other priority is impaired, there is no reason that a loan may not be made. There often is a statutory requirement that state water officials be given written notification of the loan.

## Example—Union's Plan for Augmentation

In 1975, Union Oil Company obtained court approval of a plan for augmentation allowing the conversion of existing Colorado irrigation water rights (in the Roaring Fork River and Parachute Creek Basins) for use at the company's synfuels operation on Parachute Creek. The company sought the right to divert water under the senior priorities of the changed water rights at the pumping pipeline and through Parachute Creek wells, designated as alternate points of diversion. Water diverted under the changed rights was also to be stored in Parachute Creek Reservoir.

To protect the vested interest of other water rights, the augmentation plan decree contained certain limits and conditions governing Union's exercise of its Roaring Fork and Parachute Creek rights. The following is a summary of the decree and describes the conditions of the operation of the plan for augmentation. Figure 12 shows the location of the principal elements of the Union Augmentation Plan.

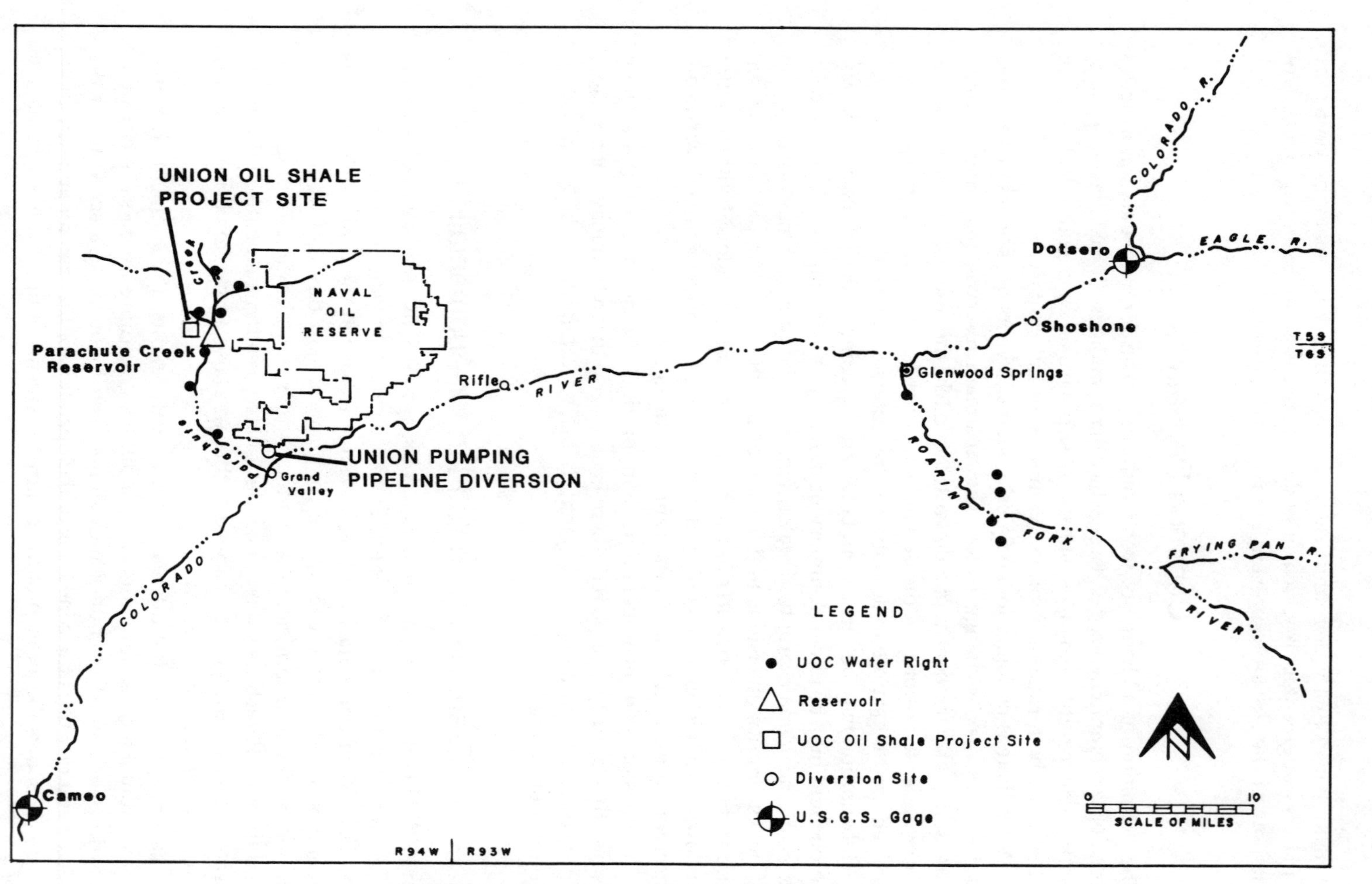


**FIGURE 12.** Union Oil Company water rights location.

1. Union's water rights are modified and can be used to supplement the water supply for Union's oil shale operation and Union is allowed to:
   a. Change the use of the Union water rights in the Roaring Fork and Parachute Creek drainages to industrial, retorting, mining, refining, power, domestic, and all other uses in connection with oil shale operations, as alternative uses when the rights are not used for irrigation.
   b. Divert under the priority of the Roaring Fork water rights at the intake of the pumping pipeline diversion facility on the Colorado River as an alternate point of diversion and to convey the water diverted by pipeline to Parachute Creek and its tributaries for use in connection with the oil shale operation as follows:
      (1) Storage in Parachute Creek Reservoir to the extent of the historic consumptive use associated with said water rights, with the right to fill and refill the reservoir continuously;
      (2) Diversion at the existing points of diversion of each of its Parachute Creek water rights and utilization in conjunction with, or substitution for, each of the Parachute Creek rights for oil shale purposes only; and
      (3) Augmentation of each of its existing wells and future wells which may be constructed in the drainage of Parachute Creek and its tributaries so that diversions through the wells shall not be curtailed in times of shortage so long as the plan for augmentation is administered in accordance with the terms of the decree.
   c. Divert, under its Parachute Creek water rights, at the following alternative points of diversions:
      (1) The existing point of diversion of each Parachute Creek water right;
      (2) Additional wells to be constructed in the drainage of Parachute Creek and its tributaries; and
      (3) The intake of its pumping pipeline diversion facility on the Colorado River.
   d. Store water in Parachute Creek Reservoir under each of its water rights in the Parachute Creek drainage to the extent of the historic consumptive use associated with those rights with the right to fill and refill the reservoir continuously. Water to be stored under the rights which have points of diversion downstream from the Parachute Creek Dam will be stored by exchange from Parachute Creek and its tributaries above the dam, or by diverting water at the intake of the pumping pipeline diversion and conveyance through the pipeline to the Parachute Creek Reservoir.
   e. Reimpound and reuse return flows, if any, from its oil shale operations, including sewage effluent, to the extent of the historic consumptive use associated with water rights in the Roaring Fork and Parachute Creek drainages after deducting carriage losses resulting from said reim-

poundment and reuse, as may be required by the division engineer for Water Division No. 5, not to exceed 0.05 percent per mile; or to the extent of the augmentation of Parachute Creek by releases from Parachute Creek Reservoir, whichever is greater.

2. The state engineer and division engineer for Water Division No. 5 are to administer the plan for augmentation including exchange in accordance with the following conditions:
   a. Diversions under Union's water rights in the Roaring Fork Drainage at the intake of the pumping pipeline on the Colorado River as an alternate point of diversion are subject to the following limitations:
      (1) Diversions can be made only during the period April 15th through September 30th;
      (2) Diversions can be made only when the Roaring Fork water rights are in priority and water is physically available at their historic points of diversion and the amounts of such diversions shall be limited to the amounts available at the historic points of diversion;
      (3) The total amount of diversions under the Roaring Fork water rights shall not exceed 1152.6 af, less the estimated carriage losses between the historic points of diversion and the intake of the pumping pipeline, as required by the division engineer, but not to exceed 29 af. During years when the Roaring Fork water rights historically used for irrigation (the historic annual consumptive use for these rights is established at 863.5 af) are diverted prior to June 30th for irrigation of the lands the amount of water which Union shall be entitled to divert at its alternate point of diversion shall be reduced by 5.1 af per day beginning on April 15th and continuing through the day on which the diversion for irrigation ceases;
      (4) During years when diversions are made only under the Roaring Fork water rights historically used to irrigate lands permanently removed from irrigation (158 acres with an average annual consumptive use of 289.1 af) and no longer owned by Union, the total of such diversions shall not exceed 289.1 af, reduced by estimated carriage losses between the historic point of diversion and the intake of the pumping pipeline, as required by the division engineer, but not to exceed 7 af. During years when the water rights are diverted prior to June 30th for irrigation of other lands owned by Union in the Roaring Fork Drainage, the amount of water diverted at the alternate point of diversion is reduced by 1.9 af per day beginning April 15th and continuing through the day on which the diversions for irrigation cease.
      (5) When diversions are made by Union under the Roaring Fork water rights, Union must place a call by the Roaring Fork water rights on the Roaring Fork River at the historic point of diversion of the rights

for an amount sufficient to produce the amount of the decreed rights. The call is to be placed only against those water rights against which a call historically could have been placed.

b. Diversions by Union's water rights in the Parachute Creek Drainage, at any of the alternate points of diversion, are subject to the following limitations:
   (1) Diversions under the Parachute Creek water rights, except for the five decreed wells and additional wells which may be constructed in the Parachute Creek Drainage, are to be made only during the period April 15th through October 31st.
   (2) Diversions can be made only when the water rights are in priority and water is physically available at the historic points of diversion and the amounts of such diversions are limited to the amounts available at the historic points of diversion;
   (3) The total consumptive use resulting from such diversions shall not exceed 1395.9 af per year;
   (4) When diverting under the Parachute Creek water rights, Union must place a call by the Parachute Creek water rights on Parachute Creek at the historic points of diversion in an amount sufficient to produce the amount of the decreed rights, but the call shall be placed only against rights against which a call historically could have been placed.
c. The storage of water in Parachute Creek Reservoir under the Union Parachute Creek water rights, which historically have been used for irrigation, is subject to the following limits:
   (1) Storage can occur only during the period April 15th through October 31st of each year;
   (2) Storage can occur only when the water rights are in priority and shall not exceed the amount of water physically available at the historic points of diversion;
   (3) The total amount of storage shall not exceed 1395.9 af in any year.
d. When water is diverted under Union's Roaring Fork water rights at the pumping pipeline alternate point of diversion for use in connection with oil shale operations, the lands owned by Union in the Roaring Fork Drainage shall be removed from irrigation with Roaring Fork water. Union is not required to remove any portion of these lands from irrigation when water is diverted at its pumping pipeline alternate point of diversion in accordance with Paragraph 2.a.(4) meaning when diversions are made only under water rights historically used to irrigate lands permanently removed from irrigation. When all of Union's Parachute Creek water rights are used in connection with oil shale operations, the Union lands in Parachute Creek must be removed from irrigation by these rights.

e. Except for water diverted in accordance with Paragraph 2.a.(4), when diversions are made only under water rights historically used to irrigate lands permanently removed from irrigation, Union shall use its water rights in the Roaring Fork Drainage in connection with its oil shale operations only when other sources of water supply for these operations are not adequate. The water rights in the Roaring Fork Drainage shall not be used to irrigate any of Union's historically irrigated lands in the Parachute Creek Drainage.

f. Union must notify the division engineer on or before June 30th of each year whether it will be necessary to use its Roaring Fork water rights to supplement its water supply for oil shale operations. Union shall not be permitted to resume the irrigation of its lands in the Roaring Fork Drainage in any year in which those water rights have been utilized in connection with oil shale operations. In any year in which Union determines that it will be necessary to use its Roaring Fork water rights to supplement its water supply for oil shale operations, all of those water rights, except water diverted in accordance with Paragraph 2.a(4), when diversions are made only under water rights historically used to irrigate lands permanently removed from irrigation, shall be used for oil shale operations. Union must notify the division engineer on or before April 15th if it will be necessary to use all or a portion of the Parachute Creek water rights to supplement its water supply for oil shale operations. If Union seeks to utilize only a portion of the Parachute Creek rights for oil shale operations, they must designate the water rights to be withdrawn from irrigation and are entitled to credit for the quantities of water set forth in Table I (copied from decree). Partial discontinuance of irrigation shall require that all lands under any single ditch identified in Table I shall not be irrigated in that year.

g. To permit withdrawals through Union's existing wells and additional wells which may be constructed in the Parachute Creek Drainage, water shall be released to Parachute Creek from Parachute Creek Reservoir, the pumping pipeline, or from other sources to the extent of withdrawals through the wells, as necessary, to meet valid calls by water rights having priority dates senior to the wells. The additional wells constructed in the Parachute Creek Drainage shall be located on patented land owned by Union. The aggregate yield from the additional wells is not to exceed 1000 gallons per minute.

h. To ensure that changes in use and points of diversion permitted by the plan of augmentation will not change the regimen of the Colorado River or its tributaries in such a way as to cause material injury to other appropriators, the decree imposes the following conditions:

(1) During years when water is diverted at the pumping pipeline alternate point of diversion under the Roaring Fork water rights, other than water diverted in accordance with Paragraph 2.4, when di-

Table I
IRRIGATED CROPS AND AVERAGE ANNUAL WATER CONSUMPTION
PARACHUTE CREEK UNION OIL COMPANY

| Ditch | Alfalfa | | Native Hay-Pasture | | Corn | | | |
|---|---|---|---|---|---|---|---|---|
| | Acres Planted | Net[a] Consumptive Use (af) | Acres Planted | Net[a] Consumptive Use (af) | Acres Planted | Net[a] Consumptive Use (af) | Total Acres Planted | Net[a] Consumptive Use (af) |
| Parachute and Vieweg | 220 | 455.4 | 0 | 0 | 10 | 14.1 | 230 | 469.5 |
| Low cost | 292 | 604.4 | 0 | 0 | 20 | 28.2 | 312 | 632.6 |
| Benson-Barnett | 0 | 0 | 2 | 3.9 | 0 | 0 | 2 | 3.9 |
| Granlee | 0 | 0 | 20 | 39.0 | 0 | 0 | 20 | 39.0 |
| East Fork | 0 | 0 | 10 | 13.0[b] | 0 | 0 | 10 | 13.0[b] |
| PLR | 0 | 0 | 10 | 13.0[b] | 0 | 0 | 10 | 13.0[b] |
| CCD | 0 | 0 | 4 | 5.2[b] | 0 | 0 | 4 | 5.2[b] |
| Charley Dere | 0 | 0 | 10 | 13.0[b] | 0 | 0 | 10 | 13.0[b] |
| Ida Dere | 0 | 0 | 12 | 15.6[b] | 0 | 0 | 12 | 15.6[b] |
| Davenport and West Fork | 0 | 0 | 98 | 191.1 | 0 | 0 | 98 | 191.1 |
| Total | 512 | 1059.8 | 166 | 293.8 | 30 | 42.3 | 708 | 1395.9 |

[a] Gross consumptive use less effective precipitation.
[b] Value represents net consumptive use less one-third to compensate for water shortages.

versions are made only under water rights historically used to irrigate lands permanently removed from irrigation, Union, during the month of October must release to the Colorado River up to 33.1 af of water;

(2) During years when Union's Parachute Creek water rights are used in connection with oil shale operations, Union must, during the month of October of such years, release to the Colorado River up to 77.4 af of water;

(3) These releases shall be required only if there are in effect during such times valid calls determined by the division engineer to result from the changes in use and points of diversions permitted by the plan of augmentation.

i. To facilitate the administration of the plan of augmentation, Union must install measuring devices approved by the division engineer for Water Division No. 5 at each point of diversion and alternate point of diversion to be utilized under the plan. Measuring devices shall be installed at the points of diversion of Union's water rights in the Roaring Fork Drainage to measure both diversions into the headgates and releases back to the river. When Union is diverting at its pumping pipeline alternate point of diversion, these diversions and releases shall be operated to adequately account for the proper amount of water needed to be physically available in the Roaring Fork Drainage below the points of diversion to allow for proper administration of the plan. The rate of flow to be diverted at the alternate point of diversion shall be no greater than the rate of flow of the water measured and returned by Union under its water rights in the Roaring Fork Drainage below their points of diversion existing as of the date of the decree.

The same provisions and conditions apply with regard to the points of diversion of Union's water rights in the Parachute Creek Drainage except that measuring devices need not be installed until the city and county of Denver or Atlantic Richfield Company, or the division engineer deem it necessary for the proper administration of the plan, provided that Union is notified at least one operating season in advance.

3. The decree states that by the imposition of the conditions contained in the decree, the operation of the plan for augmentation including exchange will not cause material injury to the owners or users of vested water rights or decreed conditional water rights in the Colorado River or its tributaries. It also decrees that, as a matter of hydrological and geological fact by virtue of the plan, there will be water available for the groundwater withdrawals contemplated in the plan, and that no material injury to vested or decreed conditional water rights in the Colorado River or its tributaries will result from the plan. Accordingly, the state engineer is directed to recognize the existence of the plan and not to deny permits for the wells included in the plan. Further, the division engineer is directed not to curtail diversions

through Union's wells as long as the plan of augmentation is being administered in accordance with the terms and conditions of the decree.

Table 3 is a summary of the Union Oil Company augmentation plan water rights and operating conditions.

Table 3
SUMMARY OF UNION OIL COMPANY AUGMENTATION PLAN (W-2206)

| Source | Amount Decreed | Maximum Diversion Per Decree in W-2206 | Conditions and Limitations |
|---|---|---|---|
| Pumping Pipeline conditional decree from Colorado River | 118.5 cfs | — | Limited only by physical supply and priority. Alternate point of diversion (APD) for Parachute Creek and Roaring Fork rights. |
| Parachute Creek Reservoir | 33,773 af | — | Initial construction proposed for 3500 af active capacity. Can store water diverted under Roaring Fork and Parachute Creek rights. |
| Parachute Creek direct flow | 95.902 cfs | 1,395.9 af | Alternate points of diversion include existing headgates, wells, and pumping pipeline. 1. Diversions limited to April 15th–October 31st and when in priority and water physically available at historic POD. Diversions through wells as APD not limited to April 15th–October 31st period. 2. Storage in Parachute Creek reservoir only April 15th–October 31st when in priority and water available at historic POD. 3. Storage cannot exceed 1395.9 af 4. Historic irrigated land must be removed from irrigation 5. Must release 77.4 af to Colorado River in October if valid call in effect |
| Wells | 0.6342 cfs plus additional 1,000 gpm | | Alternate point of diversion for direct flow rights 1. Diversion through wells not limited in time 2. Out of priority depletions must be replaced by releases from Parachute Creek reservoir or pumping pipeline |

Table 3 (*continued*)

| Source | Amount Decreed | Maximum Diversion Per Decree in W-2206 | Conditions and Limitations |
|---|---|---|---|
| Roaring Fork | 54.153 cfs | 1,152.6 af | APD at pumping pipeline. Storage in Parachute Creek reservoir and augmentation of wells.<br>1. Limited to April 15th through September 30th and when in priority and water physically available at historic POD.<br>2. Amount diverted reduced by carriage loss up to 29 af<br>3. Diversions up to 863.5 af subject to reduction of 5.1 af/day if started prior to June 30th for irrigation<br>4. Diversion up to 289.1 af for land permanently removed from irrigation reduced by carriage loss up to 7 af and reduced by 1.9 af/day if started prior to June 30th.<br>5. Must release up to 33.1 af to Colorado River in October if valid call in effect. |

*Source:* Decree in the matter of the Application for Water Rights for Union Oil Company of California, Case W-2206, District Court in and for Water Division No. 5, July 14, 1975.

# 3

# INSTITUTIONAL CONSIDERATIONS

Water rights are no exception to the pervasive influence of governments in the lives of their citizens. Whether perfecting a water right, looking for information about water rights, or enforcing water rights, one must deal with any number of governmental entities, especially their executive (administrative) and judicial branches.

## ADMINISTRATIVE SYSTEMS

The western United States has reduced the emphasis on litigation and gunplay in the water rights area by creating strong administrative systems with respect to perfection of water rights, collection of data on water and water rights, and the administration of perfected water rights. In most of the western states, all three administrative functions have been centralized in one office, usually the state engineer.

### Control and Enforcement

The state (either through administrative or judicial entities) exercises virtually plenary control over the perfection of water rights. Once water rights are perfected and put into use, the state maintains control through the process of administration. The state engineer acts as the high sheriff for water matters in the state. He has several deputies, usually called division engineers or division superintendents, who have authority over specific areas. The division engineers, in turn, employ water commissioners, cops on the beat, who exercise their authority within one or more small watersheds.

The water commissioners serve two valuable functions, regulation of priorities and collection of data. In most states, the water commissioners are responsible for collecting information on each water right, including such things as daily diversions and acres irrigated. If a dispute arises, the water commissioners are called upon to initially resolve the dispute by enforcing the priority system. On some streams which are perennially water short, they may open and shut headgates on their volition in order to avoid disputes.

## Control Through Permits and Regulations

In addition to the control exercised in many states by the permit process for the protection of water rights, in virtually every state control is exercised through permits and rules and regulations. The most far-reaching permit control exercised by various states concerns construction of wells. Almost every state requires well drillers to be licensed. If a licensee drills an unpermitted well, he is likely to lose his license. Permits for the construction of wells are required and obtained in virtually every western state. In a few states, the construction permits are issued only after considerations of public health and safety are satisfied. In an increasing number of other states, permits are issued only if there is unappropriated groundwater available and the construction and operation of the new well will not adversely affect the owners of other water rights. From state to state, permits may also be required for other water-related activities such as weather modification.

Rules and regulations are the natural companion to permits. These may govern such things as the design and construction of reservoirs and the operation of wells. As an example, consider the following rules and regulations promulgated by the state engineer for the operation of wells along Colorado's South Platte River. Although Colorado is a mandate state, under which water rights are usually perfected by court decrees, it appears the following regulations are a successful attempt by the state engineer to circumvent the mandate system.

| | | |
|---|---|---|
| IN THE MATTER OF THE RULES AND<br>REGULATIONS GOVERNING THE USE,<br>CONTROL, AND PROTECTION OF<br>SURFACE AND GROUND WATER RIGHTS<br>LOCATED IN THE SOUTH PLATTE<br>RIVER AND ITS TRIBUTARIES | )<br>)<br>)<br>)<br>)<br>) | AMENDED RULES AND REGULATIONS<br>OF THE<br>STATE ENGINEER |

Pursuant to authority vested in the Office of the State Engineer, the State Engineer hereby,

FINDS, that on November 16, 1972 the State Engineer ordered that Rules and Regulations for the South Platte River were to become effective on February 19, 1973. As a result of protests filed to those Rules and Regulations and upon the basis of subsequent proceedings in the Water Court for Water Division I, those Rules and Regulations are hereby amended and changed to read as reproduced below.

The said Amended Rules and Regulations are adopted and shall become effective as of the 16th day of March, A.D., 1974, and shall remain in full force and effect unless changed or amended as provided for by law.

"AMENDED RULES AND REGULATIONS"

RULE 1. Except as specifically noted below, these Rules and Regulations shall apply to all underground water of the South Platte River and its tributaries as defined in *Colo. Rev. Stat. Ann.* 1963, Sec. 148-21-3(4) (Supp. 1969) 37-92-103, CRS 1973 , and reproduced below, as follows:

> (4) "Underground water" as applied in this act for the purpose of defining the waters of a natural stream, means that water in the unconsolidated alluvial aquifer of sand, gravel, and other sedimentary materials, and all other waters hydraulically connected thereto which can influence the rate of direction of movement of the water in that alluvial aquifer or natural stream. Such "underground water" is considered different from "designated ground water" as defined in 148-18-2(3) 37-90-103, CRS 1973 .

These Rules and Regulations shall not apply to water withdrawn from wells, such as domestic and livestock wells, which are exempted from administration under *Colo. Rev. Stat. Ann.* 1963, Section 148-21-45 (Supp. 1972) 37-92-602, CRS 1973 , and these Rules and Regulations shall not apply to water withdrawn from wells which are exempted from administration by Court decree or statute.

RULE 2. (a) Ground water diversions will be continuously curtailed according to the following schedule to provide for a reasonable lessening of material injury to senior appropriators:

(1) During the Calendar Year 1974, five-sevenths (5/7) of the time;

(2) During the Calendar Year 1975, six-sevenths (6/7) of the time; and

(3) During the Calendar Year 1976, and thereafter, total curtailment.

Pumping shall be permitted on every Monday and Tuesday of each week in 1974 and on every Monday of each week in 1975. The Division Engineer shall administer this rule so that the operator of a well, or wells, may have a cycle of operation to make more efficient use of the water available; provided, that senior appropriators are not materially injured thereby.

(b) Ground water diversions shall be curtailed as provided under part (a) hereof unless the ground water appropriator submits proof to the Division Engineer and upon the basis of that proof the Division Engineer shall find:

(1) That the well is operating pursuant to a decreed plan of augmentation, that the well is operating pursuant to a decree as an alternate point of diversion, or that a change in point of diversion to the well has been decreed for a surface water right; or

(2) That the ground water appropriation can be operated under its priority without impairing the water supply to which a senior appropriator is entitled, or

(3) That the water produced by a well does not come within the definition of underground water in RULE 1.

RULE 3. Any ground water appropriator affected by these Rules and Regulations may use a part or all of the water diverted without regard to curtailment described in RULE 2(a) to the extent his ground water diversion is in compliance with a temporary augmentation plan approved by the Division Engineer in accordance with Colo. Rev. Stat. Ann. 1963, Sec. 148-21-23(4) 37-92-307, CRS 1973 and where there is a plan for augmentation filed in the Water Court in Accordance with Colo. Rev. Stat. Ann. 1963, Sec. 148-21-18 (Supp. 1971) 37-92-302, CRS 1973 . The Division Engineer will promptly approve or disapprove such temporary augmentation plans submitted to him. The guidelines for any such temporary augmentation plan will be expected to meet at least the following criteria:

(1) That replacement water for stream depletion shall be made available to the Division Engineer in an amount equal to 5 percent of the projected annual volume of a ground water diversion, and may be used by him at a rate of flow sufficient to compensate for any adverse effect of such ground water diversion on a lawful senior requirement, as evidenced by a valid senior call, but at a rate not exceeding 5% of the capacity of the diversion structure.

(2) Such capacity shall be determined by Court decree, if adjudicated, by application for a water right, if filed in the Water Court, by well permit, or by registration. If none of these means of determination

is available, the capacity will be the maximum pumping or delivery rate, which must be substantiated by the appropriator.

(3) The operation of the temporary augmentation plan shall not be used to allow ground water withdrawal which would deprive senior surface rights of the amount of water to which said surface rights would have been entitled in the absence of such ground water withdrawal, and ground water diversions shall not be curtailed nor required to replace water withdrawn, for the benefit of surface right priorities, even though such surface right priorities be senior in priority date, when, assuming the absence of ground water withdrawal by junior priorities, water would not have been available for diversion by such surface right under the priority system.

RULE 4. Whenever the Division Engineer is satisfied, upon the basis of competent evidence, that operation of a temporary plan of augmentation pursuant to RULE 3(1) will not meet the requirements of RULE 3(3) above, modification of the plan will be undertaken by reference to criteria as follows:

(1) The stream depletion caused by a well will be calculated by the method shown in The Pumped Well by Robert E. Glover, Technical Bulletin 100, Colorado State University or by other accepted engineering formulae appropriately modified to reflect the pertinent physical conditions.

(2) The transmissivity value will be obtained from the U. S. Geological Survey Open-File Reports, Hydrogeologic Characteristics of the Valley-Fill Aquifer in the South Platte River Valley, Colorado, 1972, or from updated editions, or from cal ulations using accepted engineering methods.

(3) The specific yield or effective voids ratio generally descriptive of the material in the aquifer will be assumed to be twenty percent (20%), or a different value may be used when it can be substantiated generally or as to any particular area or situation.

(4) The consumptive use for irrigation purposes will be assumed to be forty percent (40%) of the total quantity pumped for irrigation uses, subject to modification upon proof that a different consumptive use situation exists with respect to a particular diversion. For uses other than irrigation, the amount will be determined from the actual conditions.

Dated this 15th day of March, 1974.

/s/ C. J. Kuiper
C. J. Kuiper, State Engineer

## Water Rights Records

### *Need for Records*

Accurate and complete records documenting water rights proceedings and administration and water system operation are an essential part of the water resource system. Records of judicial proceedings document the appropriative process and serve as a mandate to state water officials for the administering of priorities. Administrative records define the historic operation of water rights, which when used with hydrologic records provide the basis for defining the factual situation, in connection with the adjudication of new rights and the conversion of adjudicated rights to new types, places, and time of use.

Water right records facilitate water resource planning by defining the physical and legal supply available to proposed projects. Without accurate records of runoff, diversions, and uses, it is difficult to evaluate the quantity and timing of available water supplies and establish the feasibility of new water development projects.

Processing of water rights applications depends on the availability of records which establish the amount of water available to claims for new appropriations and evaluate the impact of water right changes. It is essential that water rights be quantified in terms of decreed amounts. Unquantified claims, such as those exerted by the federal government under the reserved rights doctrine, impede the full development of state water resources as long as the timing and amount of future demands, by the United States, under senior priorities, are unknown and thus cannot be accounted for in the quantification of existing demands.

The extent, accuracy, availability, and method of organization and accessibility of water rights records vary from state to state. The extent and accuracy of records increases as the demand for water increases on a particular stream system, as well as the need for stricter administration of the scarce supplies and a better accounting of uses.

### *Statutory Requirements*

Usually the legislature by statute requires the state engineer to maintain virtually every conceivable record concerning water and water rights. The problem is that the state engineer is seldom given enough money by the legislature to adequately fulfill his statutory responsibilities. There are sometimes lapses in the collection of basic data documents, but the more common lapse is the accessibility of those materials. Although state engineers often welcome and encourage public use of their files, the indexing systems, personal assistance, and physical availability of the files often are less than ideal.

In most states, there is a statutory provision which gives great weight to the records of the state engineer. They are usually prima facie evidence of the truth of their contents in a court proceeding. While this provision may be convenient, it is often far from satisfactory. The state engineer's records are often the best and, perhaps, the only records available. Nevertheless, they are often inaccurate.

This is particularly true with respect to the diversions or amount of water stored by a particular water right. In water districts where there has been little controversy, the water commissioner usually has not played an active role. The headgate or reservoir may be visited only two or three times a year. In fact, they may never be visited. Regardless of the number of visits made, the water commissioner usually relies on the statements of the water right owner to develop records of water use. If the water commissioner does not speak with the owner, he may simply assume the water right has diverted its full perfected amount each day of the irrigation season, unless he has made a few visits to the headgate and recorded measurements. In that case, he will assume that the diversions being made at one of his visits continue, without fluctuation, until his next visit. In water districts where controversy is prevalent, the water commissioners make frequent visits and, based on their own personal observations, collect reliable data.

## Types of Records

Water rights records can be broadly classified as documental, operational, and hydrologic, depending primarily on the source, nature, and use of the records. The various types of water rights records kept by states operating under the appropriation system are summarized below:

*Documental.* These are records of judicial, administrative, and other proceedings which document the factual evidence, legal principles, and final results which provide a basis for future administration.

1. Court decrees adjudicating absolute and conditional water rights, changes, transfers, and modifications of water rights, plans of augmentation, and other judicial proceedings.
2. Interstate compacts establishing the basis for allocation of water from interstate streams.
3. Administrative proceeding records documenting the nature and results of proceedings before administrative hearing officers, boards, or commissions.

*Operational.* These records that are collected and maintained by the administrative office measure the operation of the water rights system. Basic data generally are supplemented by additional information, depending on the system.

1. Common data
   a. Location
      (1) Water division
      (2) Water district
      (3) Section, Township, and range
      (4) County
2. Ditch diversions
   a. Structure name/Owner/Priority
   b. Source
   c. Amount
   d. Use
   e. Area irrigated
   f. First and last day of diversion

3. Reservoir storage
   a. Structure name/Owner/Priority
   b. Use
   c. Capacity
   d. Source
   e. Elevation/Area
   f. Drainage area
   g. Gage height
   h. Inflow/Outflow
   i. Dam
      (1) Height
      (2) Spillway description
      (3) Type of dam
      (4) Outlet works
4. Water rights data from decrees
   a. Structure name
   b. Source
   c. Quantity
   d. Appropriation date
   e. Adjudication date
   f. Use
   g. Priority/Basin rank
5. Stock ponds
   a. Structure name
   b. Drainage area
   c. Capacity
   d. Surface area at high water
   e. Dam height
   f. Outlet size
6. Wells
   a. Permit number
   b. Use
   c. Yield
   d. Depth
   e. Annual appropriation
   f. Aquifer
   g. Drilling log
   h. Casing data
   i. Completion/Test data
   j. Annual withdrawal
   k. Water level (date and depth below measuring point)

*Hydrologic.* These records measure the climate and hydrology, including, precipitation and runoff. Collection and maintainence is by cooperative effort between the state and federal water and climatological agencies.

1. Gaging stations
   a. Identification
   b. Stream or ditch
   c. Period of record
   d. Elevation
   e. Drainage area
   f. Extremes of records
   g. Nature of records
   h. Daily flow
2. Climate
   a. Station
   b. Location
   c. Elevation
   d. Period of record
   e. Precipitation
   f. Snow depth/Water content
   g. Wind direction/Velocity
   h. Evaporation
   i. Temperature
   j. Solar radiation

## *Use of Records*

Records of water rights proceedings and operation provide the basic data needed for defining the factual basis for judicial decisions that protect vested water rights from injury in matters involving the adjudication of new rights and the conversion of decreed rights to new types, times, and places of use. To ensure that vested rights are not injured, it is necessary to define historic conditions on a stream system in terms of seasonal and annual flow and the diversion history of the ditches involved. It is also necessary to establish the historic burden placed on the stream system by the ditches, through computation of the consumptive use utilizing climatic data, including temperature, precipitation, and solar radiation. Once the historic operation of the stream system has been defined, the impact of changing water rights or appropriating new water rights can be evaluated. This can be done by superimposing the proposed new conditions on the historic operation and modifying the proposed new conditions to ensure that the use of the water rights is not enlarged in terms of the period and amount diverted and depleted from the surface stream system.

Various formats are used for recording water rights data. The following three examples illustrate the tabulation of water rights by the state of Wyoming, the Colorado tabulation of decreed rights by stream, which includes a list of rights considered abandoned and a sample of a field book record of diversions as maintained by a water commissioner in Colorado. In recent years (essentially since passage of the 1969 Water Rights Determination and Administration Act), Colorado has attempted to computerize the tabulation of water rights decrees and diversion records.

## WYOMING

### EXPLANATION OF TABLES

USE:

| | | | | |
|---|---|---|---|---|
| Chem.—Chemical | Eng.—Steam Engines | Ind.----Industrial | Misc.—Miscellaneous | Rec.----Recreational |
| Com.---Commercial | Fire—Fire Protection | Mech.---Mechanical | Mun.---Municipal | Ref.----Refining |
| Cul.---Culinary | Fish—Fish Propagation | Mfg.----Manufacturing | Oil----Oil Refining or Production | S-------Stock |
| D------Domestic | F.C.—Flood Control | Mil.----Milling | Power—Power Development | Trans.—Transportation |
| Drl.---Drilling | I-----Irrigation | Min.----Mining | RR.----Railroad | |

### SURFACE WATER

ARRANGEMENT OF TABLES:

First----for the main stream. Parenthesized numbers following a source name indicate either the mouth of a stream or the location of a spring or springs.
Second---for tributaries complete, beginning at the mouth of the stream or at the state boundary line and proceeding toward the source of such main stream.
Third----on each stream according to date of priority of the appropriation.
The rights are shown under heading of the stream from which diversion is actually made, according to the best information available, although in some cases the source of supply was given in the early adjudication as the main stream or the name of the tributary was erroneously given.

COLUMN HEADINGS:

PERMIT

Terr.—Territorial Appropriation
592 R--Permit No. 592 Reservoir
1398 E--Permit No. 1398 Enlargement
10 SR--Permit No. 10 Stock Reservoir

Court Decree priorities are indicated by number.
Where Permits do not exist, Proof numbers are given, or the column has been left blank.
Double Permits (5686-6684) — The latter is an amendment of the first, before petitions were required.

Priority (Abbreviations for):
Month, day and year-----2-13-1892
Month and year----------2-00-1892
Year--------------------- -1889
Summer 1889-------------Sum.-1889
Spring 1885-------------Spg.-1885
Season 1884-------------Sea.-1884

Order of Administration of Territorial Priorities
Day, month and year--------May 15, 1884
Month and year------------- May, 1884
Specified Season and year—Spring, 1884
Season and year------------Season, 1884
Year only------------------ 1884
Before year----------------Before 1885

Board Orders or Court Orders might also establish a specific priority.

C.F.S. (Amount of Appropriation)
c.f.s.----------Cubic Feet per Second, for ditches.
a.f.------------Acre Feet, for reservoirs.
S.S.------------Supplemental Supply, for lands having an original supply from another source.
Sec.Sup.--------Secondary Supply, for water stored in a reservoir.
Supply Ditch----Supplies water to a reservoir.

H.G.Loc. Headgate Location or Reservoir Outlet or Spillway, given by Section, Township, and Range — 30-55-101.
S.T.R. Resurvey is indicated by Tr. (Tract)— Tr.69-33-90, or L. (Lot) — L.4-33-90.

UNIT OF MEASURE: Wyoming law provides that measurement of flowing water shall be in terms of cubic feet per second of time. The volume of stored water is measured in acre-feet.

| | | | |
|---|---|---|---|
| 1 cu. ft. | = | 7.48 | gallons, which weigh 62.4 pounds |
| 1 cu. ft. per sec. | = | 448.83 | gallons per minute. |
| | = | 26,930.00 | gallons per hour. |
| | = | 646,317.00 | gallons per day. |
| | = | 1.9835 | acre-feet per day, or for practical purposes, 2 acre-feet per day is used for administration of small heads of water. |
| 1 acre-foot | = | 43,560.00 | cubic feet, or the quantity necessary to cover one acre one foot in depth. |

### GROUND WATER

"Ground Water" or "Underground Water" — These terms are used interchangeably and by law mean any water, including hot water and geothermal steam, under the surface of the land or the bed of any stream, lake, reservoir, or other body of surface water, including water that has been exposed to the surface by an excavation such as a pit.

ARRANGEMENT OF TABLES:

First----by district number.
Second---alphabetically by county name.
Third----by control area (if any).
Fourth---by sub-district (if any).
Fifth----according to date of priority of the appropriation.

COLUMN HEADINGS:

U.W. No. (Underground Water Number)
SC------Statement of Claim, a term applied to rights established for wells constructed prior to April 1, 1947, where the "Statement of Claim" was properly filed prior to March 1, 1958.
WR------Well Registration, a term applied to rights established for wells constructed after April 1, 1947, but prior to March 1, 1958, where the "Well Registration" was properly filed prior to March 1, 1950.
P-------Permit, a term applied to rights established for wells which have been properly filed since March 1, 1958, and do not qualify as Statements of Claim or Well Registrations.

Priority date for:
Statement of Claims is the date of completion of the well.
Well Registrations is the date of filing in the State Engineer's Office.
Stock and/or domestic wells completed prior to May 24, 1969, and filed on or before December 31, 1972, is the date of completion of the well.
Stock and/or domestic wells completed prior to May 24, 1969, and filed after December 31, 1972, is the date of filing.
All other ground water rights is the date of filing.

G.P.M. (Amount of Appropriation)
G.P.M.---------Gallons Per Minute.
Add'l Supply---Additional Supply, for lands having an original supply from another source.

Well Loc. Well Location, given by Section, Township, and Range — 30-55-101.
S.T.R. Resurvey is indicated by Tr. (Tract) — Tr.69-33-90, or L. (Lot) — L.4-33-90.

OTHER RELATED TERMS:

Commingle--------Refers to water from two or more sources which is conveyed through a common distribution system.

WYOMING

1

WATER DIVISION NUMBER ONE

### NIOBRARA RIVER OR RUNNING WATER CREEK, Tributary of Missouri River

| PERMIT | DITCH | APPROPRIATOR | PRIORITY | USE | C. F. S. | ACRES | H. G. LOC S. T. R. | PRESENT OWNER |
|---|---|---|---|---|---|---|---|---|
| Terr. | E. B. Wilson | Eugene B. Wilson | 5-01-1881 | I,S | 5.85 | 400.00 | 5-32-64 | |
| Terr. | L. L. No. 2 | Luke Voorhees Cattle Co. | Sum. 1881 | I | 0.02 | 1.50 | 10-31-61 | |
| Terr. | Van Tassell No. 4 | R. S. Van Tassell | 4-08-1882 | | | | | |
| | (Declared abandoned, November 19, 1959.) | | | | | | | |
| Terr. | Van Tassel No. 5 | Harlan P. Zerbe, et al. | 4-12-1882 | I,S | 0.53 | 37.00 | 7-31-60 | |
| | (Amended certificate issued to successors of R. S. Van Tassell, original appropriator. Point of diversion and means of conveyance for 22 acres changed to Pump No. Five, 17-31-60, for 15 acres to Ditch No. Five, 17-31-60. Declared abandoned of 0.47 c.f.s. and 33 acres from 1 c.f.s. and 70 acres.) | | | | | | | |
| Terr. | Jenks Dams | Geo. D. Jenks | Sum. 1883 | I | 3.57 | 250.00 | 1-31-62 | |
| Terr. | Snyder | Henry C. Snyder | Sea. 1884 | | | | | |
| | (Actually diverts from Duck Creek, tributary of Niobrara River or Running Water Creek.) | | | | | | | |
| Terr. | Van Tassell No. 1 | R. S. Van Tassell | 4-05-1885 | | | | | |
| | (Actually diverts from Van Tassell Creek.) | | | | | | | |
| Terr. | Van Tassell No. 2 | R. S. Van Tassell | 4-10-1885 | | | | | |
| | (Actually diverts from Van Tassell Creek.) | | | | | | | |
| Terr. | Van Tassell No. 3 | R. S. Van Tassell | 4-20-1885 | | | | | |
| | (Actually diverts from Van Tassell Creek.) | | | | | | | |
| Terr. | Running Water No. 1 | Addison A. Spaugh | 5-05-1885 | I | 0.07 | 5.00 | 6-32-64 | |
| Terr. | Running Water No. 2 | Addison A. Spaugh | 5-05-1885 | I | 1.43 | 100.00 | 6-32-64 | |
| Terr. | Rock Draw | Addison A. Spaugh | 5-05-1885 | I | 1.14 | 80.00 | 6-32-64 | |
| Terr. | Niobrara No. 1 | Niobrara Ditch Co. | 5-00-1886 | I | 0.35 | 25.00 | 7-32-63 | |
| Terr. | L. L. No. 1 | Luke Voorhees Cattle Co. | Sum. 1887 | | | | | |
| | (Declared abandoned, November 10, 1932.) | | | | | | | |
| Terr. | Baker No. 1 | Nat Baker | Sea. 1887 | I,S | 1.00 | 70.00 | 10-32-64 | |
| Terr. | Baker No. 2 | Nat Baker | Sea. 1887 | | | | | |
| | (Actually diverts from Quinn Creek or Spring Creek, tributary of Niobrara River or Running Water Creek.) | | | | | | | |
| Terr. | Johnson | Ellis Johnson | 1887 | I | 0.17 | 12.00 | 12-32-64 | |
| Terr. | Emma | Frank A. Watt | 7-00-1891 | | | | | |
| | (Actually diverts from Silver Springs Creek or South Branch Niobrara River or Running Water Creek.) | | | | | | | |
| Terr. | Dog Town Nos. 1,2,3 | Jos. S. Hull, et al. | 1891 | I | 2.28 | 160.00 | 24-32-63 | |
| 9925 | John Pfister No. 1 | John Pfister | 7-05-1910 | I | 0.83 | 58.00 | 32-32-62 | |
| 9926 | Olsen No. 1 | John Pfister | 7-05-1910 | I | 1.03 | 72.00 | 30-32-62 | |
| 10042 | Jane Pfister | R. V. Pfister | 8-15-1910 | I | 1.01 | 71.00 | 19-32-62 | |
| 12283 | Ballengee | Midwestern Investment Co. | 3-16-1914 | I | 0.44 | 31.00 | 8-32-63 | |
| 2969E | Enl. Ballengee | S. A. Ballengee | 6-11-1914 | I | 0.36 | 25.00 | 8-32-63 | |
| 14538 | John Pfister No. 10 | Ella J. Pfister | 8-26-1916 | I | 0.49 | 34.00 | 32-32-62 | |
| | (Water is also stored in Pfister Res., unadjudicated Permit 2483R.) | | | | | | | |
| 81SR | Choke Cherry Stock Res | W. J. Tooley | 9-25-1945 | S | 1.21 a.f. | | 7-32-63 | |
| 5813R | McMaster Res | Andrew McMaster | 3-21-1951 | S | 62.80 a.f. | | 13-31-61 | |
| 20912 | McMaster Pump and Sprinkler Facility No. 1 | McMaster and Son, Inc | 10-24-1951 | I | 1.91 | 133.60 | 10-31-61 | |
| 6260R | McMaster No. 3 Res | McMaster and Son, Inc | 10-20-1955 | I,S | 7.73 a.f. | | 4-31-61 | |
| 6458R | Zerbe No. 4 Res | Harlan P. Zerbe, et al | 6-28-1957 | I,S | 16.59 a.f. | | 17-31-60 | |
| 6459R | Zerbe No. 5 Res | Harlan P. Zerbe, et al | 6-28-1957 | I,S | 17.84 a.f. | | 17-31-60 | |
| 6506R | O'Brien Irrigation and Stock Res | Joe O'Brien | 11-06-1959 | I,S | 46.45 a.f. | | 15-32-63 | |
| 22165 | Adams North | Lawrence Adams, et al | 11-04-1960 | I | 0.33 | 22.50 | 18-31-60 | |
| 22166 | Adams South | Lawrence Adams, et al | 11-04-1960 | I | 0.15 | 10.50 | 18-31-60 | |
| 22325 | Siebken | George C. Siebken, et al | 6-30-1961 | I,S | 0.84 | 59.00 | 17-31-60 | |
| 6053E | Enl. Siebken | George C. Siebken, et al | 10-22-1962 | I,S | 0.19 | 13.00 | 17-31-60 | |
| 22622 | McMaster No. 1 | McMaster and Son, Inc | 4-07-1964 | I | 26.15 a.f. Sec. Sup. | 133.20 | 4-31-61 | |
| | (Secondary supply stored in McMaster No. 3 Res., Permit 6260R. and Enl. McMaster No. 3 Res. Permit 6822R. Original supply is through the McMaster Pump and Sprinkler Facility No. 1, Permit 20912.) | | | | | | | |
| 6822R | 1st Enl. McMaster No. 3 Res | McMaster and Son, Inc | 4-07-1964 | I | 18.42 a.f. | | 4-31-61 | |

### NIOBRARA RIVER OR RUNNING WATER CREEK and QUINN CREEK or SPRING CREEK, Tributary Niobrara River or Running Water Creek

| PERMIT | DITCH | APPROPRIATOR | PRIORITY | USE | C. F. S. | ACRES | H. G. LOC S. T. R. | PRESENT OWNER |
|---|---|---|---|---|---|---|---|---|
| 4646R | Golf Course Res | Niobrara Country Club | 3-20-1937 | S,D,Fish | 4.50 a.f. | | 12-32-64 | |

### VAN TASSELL CREEK, Tributary Niobrara River or Running Water Creek

| PERMIT | DITCH | APPROPRIATOR | PRIORITY | USE | C. F. S. | ACRES | H. G. LOC S. T. R. | PRESENT OWNER |
|---|---|---|---|---|---|---|---|---|
| Terr. | Van Tassell No. 1 | R. S. Van Tassell | 4-05-1885 | I,S | 1.71 | 120.00 | 9-31-60 | |
| | (Originally adjudicated from Niobrara River or Running Water Creek.) | | | | | | | |
| Terr. | Van Tassell No. 2 | R. S. Van Tassell | 4-10-1885 | I,S | 0.57 | 40.00 | 16-31-60 | |
| | (Originally adjudicated from Niobrara River or Running Water Creek.) | | | | | | | |
| Terr. | Van Tassell No. 3 | R. S. Van Tassell | 4-20-1885 | I,S | 1.14 | 80.00 | 16-31-60 | |
| | (Originally adjudicated from Niobrara River or Running Water Creek.) | | | | | | | |

### DUCK CREEK, Tributary Niobrara River or Running Water Creek

| PERMIT | DITCH | APPROPRIATOR | PRIORITY | USE | C. F. S. | ACRES | H. G. LOC S. T. R. | PRESENT OWNER |
|---|---|---|---|---|---|---|---|---|
| Terr. | Snyder | Henry C. Snyder | Sea. 1884 | I | 0.07 | 5.00 | 10-31-61 | |
| | (Originally adjudicated from Niobrara River or Running Water Creek.) | | | | | | | |
| 18121 | Christian | J. W. Christian | 4-07-1932 | I,S | 1.29 | 90.00 | 4-31-61 | |
| 18122 | Christian | J. W. Christian | 4-07-1932 | I,S | 54.20 a.f. Sec. Sup. | 90.00 | 4-31-61 | |
| | (Water is stored in Christian Res., Permit 4481R.) | | | | | | | |
| 4481R | Christian Res | J. W. Christian | 5-04-1932 | I,S,D | 54.20 a.f. | | 4-31-61 | |
| 19425 | Mill | George Mill | 8-09-1940 | I | 1.55 | 108.50 | 33-33-63 | |

TABULATION OF ADJUDICATED RIGHTS — WYOMING

| PERMIT | DITCH | APPROPRIATOR | PRIORITY | USE | C. F. S. | ACRES | H. G. LOC S. T. R. | PRESENT OWNER |
|---|---|---|---|---|---|---|---|---|
| | **SILVER SPRINGS CREEK OR SOUTH BRANCH NIOBRARA RIVER OR RUNNING WATER CREEK, Tributary Niobrara River or Running Water Creek** | | | | | | | |
| Terr. | Emma | Frank A. Watt | 7-00-1891 | I,S | 0.07 | 5.00 | 19-32-63 | |
| | (Adjudicated as from Niobrara River or Running Water Creek.) | | | | | | | |
| 9676 | Reynolds No. 1 | Estella L. Reynolds | 3-28-1910 | I | S.S. | 32.00 | 19-32-63 | |
| | (Original supply is from North Branch Silver Springs Creek or North Branch Running Water Creek, through Reynolds No. 2 Ditch, Permit 9677. Change in point of diversion from 20-32-63.) | | | | | | | |
| 14166 | Mashek No. 1 | Alex Mashek, et al | 4-22-1916 | I,S | 1.03 | 72.30 | 27-32-63 | |
| 14167 | Mashek No. 2 | Alex Mashek, et al | 4-22-1916 | I,S | 0.19 | 13.50 | 27-32-63 | |
| 14168 | Mashek No. 3 | Alex Mashek, et al | 4-22-1916 | I,S | 0.85 | 59.90 | 27-32-63 | |
| 3300R | Mashek No. 1 Res | Alex Mashek, et al | 4-22-1916 | I,S | 4.12 a.f. | | 27-32-63 | |
| 18345 | Hunter No. 1 | Myrtle Hunter, et al | 10-10-1933 | I | 0.45 | 32.00 | 23-32-63 | |
| 18346 | Hunter No. 2 | Myrtle Hunter | 10-10-1933 | I | 0.16 | 11.00 | 23-32-63 | |
| 18347 | Hunter No. 3 | Olive A. Hunter | 10-10-1933 | I | 0.07 | 5.00 | 23-32-63 | |
| 18348 | Hunter No. 4 | Olive A. Hunter | 10-10-1933 | I | 0.07 | 5.00 | 23-32-63 | |
| 20193 | Kilmer | V. C. Kilmer | 8-06-1948 | I | 0.44 | 31.00 | 19-32-63 | |
| | (Adjudicated as from Spring Creek.) | | | | | | | |
| 6553R | Hoblit Res | Annabelle Hoblit, et al | 11-28-1960 | I,S,Fish | 48.60 a.f. | | 19-32-63 | |
| 7147R | Kilmer No. 1 Res | Venus Kilmer | 8-25-1966 | I | 7.59 a.f. | | 19-32-63 | |
| 6426E | Enl. Reynolds No. 1 | Annabelle Hoblit, et al | 11-02-1971 | I | S.S. | 14.40 | 19-32-63 | |
| | (Original supply is from North Branch Silver Springs Creek or North Branch Niobrara River or Running Water Creek through Enl. Reynolds No. 2 Ditch, Permit 6427E.) | | | | | | | |
| 7451R | Enl. Hoblit Res | Annabelle Hoblit, et al | 11-02-1971 | I | 11.07 a.f. | | 19-32-63 | |
| | **SILVER SPRINGS, Tributary Silver Springs Creek or South Branch Niobrara River or Running Water Creek** | | | | | | | |
| 1948 | Silver Springs | William Reynolds | 8-29-1898 | I,S | 0.42 | 30.00 | 14-31-64 | |
| 653E | Enl. Silver Springs | William Reynolds | 5-10-1901 | I,S | 0.11 | 8.00 | 15-31-64 | |
| | (Adjudicated as from Silver Springs Creek. Erroneously adjudicated as Martha Reynolds Ext. Ditch.) | | | | | | | |
| 7054 | Silver Springs No. 1 | William Reynolds | 1-13-1906 | I | 0.28 | 20.00 | 11-31-64 | |
| 802R | Silver Springs Res | William Reynolds | 1-13-1906 | I | 12.30 a.f. | | 11-31-64 | |
| | (Stored water is for Silver Springs Ditch, Permit 1948.) | | | | | | | |
| | **PAGE FLAT DRAW, Tributary Silver Springs** | | | | | | | |
| 1311SR | Fosher No. 2 Stock Res | Harold Fosher | 12-06-1955 | S | 3.70 a.f. | | 1-31-65 | |
| 1316SR | Fosher No. 1 Stock Res | Harold Fosher | 12-08-1955 | S | 10.98 a.f. | | 1-31-65 | |
| | **DRY DRAW, Tributary Page Flat Draw** | | | | | | | |
| 4677R | Wilson Res | Isabel M. Wilson | 5-26-1937 | S | 7.27 a.f. | | 29-32-64 | |
| | **NORTH BRANCH SILVER SPRINGS CREEK OR NORTH BRANCH NIOBRARA RIVER OR RUNNING WATER CREEK, Tributary Silver Springs Creek or South Branch Niobrara River or Running Water Creek** | | | | | | | |
| 9677 | Reynolds No. 2 | Estella L. Reynolds | 3-28-1910 | I | 0.45 | 32.00 | 20-32-63 | |
| | (Adjudicated as from North Branch Niobrara River or Running Water Creek.) | | | | | | | |
| 6427E | Enl. Reynolds No. 2 | Annabelle Hoblit, et al | 11-02-1971 | I | 0.21 | 14.40 | 20-32-63 | |
| | **REYNOLDS SPRING, Tributary North Branch Silver Springs Creek or North Branch Niobrara River or Running Water Creek** | | | | | | | |
| 22182 | Hydraulic Ram Water System | Annabelle Hoblit, et al | 11-28-1960 | I,D | 0.04 | 1.00 | 19-32-63 | |
| | **QUINN CREEK OR SPRING CREEK, Tributary Niobrara River or Running Water Creek** | | | | | | | |
| Terr. | Baker No. 2 | Nat Baker | Sea. 1887 | I,S | 0.43 | 30.00 | 10-32-64 | |
| 20457 | Bredthauer No. One | W. T. Bredthauer | 6-14-1949 | I,S | 0.93 | 65.20 | 11-32-64 | |
| 5734R | Quinn No. One Res | W. T. Bredthauer | 6-14-1949 | I | 40.25 a.f. | | 11-32-64 | |
| 20458 | Bredthauer No. One | W. T. Bredthauer | 8-3-1950 | I | 40.25 a.f. Sec. Sup. | 65.20 | 11-32-64 | |
| | (Water is stored in Quinn No. 1 Res., Permit 5734R.) | | | | | | | |
| 6372R | First Enl. Quinn No. One Res | W. T. Bredthauer | 7-02-1956 | I,G | 102.65 a.f. | | 11-32-64 | |
| | **QUINN SPRING, Tributary Quinn Creek or Spring Creek** | | | | | | | |
| 15381 | | W. C. Irvine | 2-18-1919 | S,D | 0.10 | | 3-32-64 | |
| | **SPRING BRANCH DRAW, Tributary Niobrara River or Running Water Creek** | | | | | | | |
| 7317R | Spring Branch Res | Running Water Ranch, Inc | 6-29-1970 | I,S | 42.70 a.f. | | 4-32-64 | |
| | **TRIBBLE DRAW, Tributary Niobrara River or Running Water Creek** | | | | | | | |
| 1702SR | Tribbe No. 1 Stock Res | Harold Fosher | 11-23-1956 | S | 13.40 a.f. | | 35-32-65 | |

CONCERNING THE TABULATION LISTING ALL)
DECREED WATER RIGHTS AND CONDITIONAL )     WATER RIGHTS TABULATION
WATER RIGHTS IN WATER DIVISION 6       )

Notice is hereby given that pursuant to Section 37-92-402, C.R.S. (1973 & 1983 supp.) the Division Engineer of Water Division No. 6, in consultation with the State Engineer, has made such revisions as determined to be necessary or advisable to the July 1, 1978, tabulation of all decreed water rights and conditional water rights in the Water Division, including the omission or modification of water rights which he has determined to have been abandoned in whole or in part. The 1984 tabulations also refect the judgments and decrees of the Division Water Judge entered prior to January 1, 1984. The tabulation may not include certain water rights claimed in proceedings not concluded under Section 37-92-601, C.R.S. 1973.

A separate supplement to this tabulation, listing the decreed water rights which the Division Engineer has determined to be abandoned in whole or in part may be inspected in these same offices and at the same times. The Division Engineer will furnish or mail a copy of the abandonment list to anyone requesting the same upon payment of a fee of five dollars ($5).

Separate priority lists of decreed water rights and conditional water rights which take or will take water from the same source and are in a position to affect one another will be on the same priority list and will be available on request and payment of the cost to the State of Colorado.

The tabulation may be used by the Division Engineer, the State Engineer and their staffs for administrative purposes. The listing of the water rights in the tabulation shall not create any presumption against abandonment and the relative listing of water rights in the tabulations shall not create any presumption of seniority. The tabulation shall not be construed to modify special provisions of court decrees adjudicating, changing, or otherwise affecting such water rights or to modify contractual arrangements governing the interrelationship of such water rights. The tabulation may be inspected after July 1, 1984, in the offices of the Division Engineer, each Water Commissioner and each County Clerk and Recorder at any time during the regular office hours. The Division Engineer will furnish or mail a copy of the tabulation to anyone requesting the same upon a payment of a fee of five dollars ($5).

Dated this 1st day of July 1984.

Wesley E. Signs, Division Engineer
Jeris A. Danielson, State Engineer

---

EXPLANATION OF CODES USED IN THE WATER RIGHTS TABULATION

COLUMN (1) - THIS COLUMN INDICATES THE TYPE OF STRUCTURE AND IS CODED AS FOLLOWS:
D DITCH   R RESERVOIR   SE SEEPS   PL PIPELINE   PP POWER PLANT
W WELL   SP SPRING   M MINE   P SURFACE PUMP   O OTHER
(*=COMBINATION OF TWO OR MORE OF THE PRECEDING)

COLUMN (WD) - THE NUMBER SIGNIFIES THE FORMER WATER DISTRICT.
THE FOLLOWING EIGHT EXCEPTIONS SIGNIFY PORTIONS OF THE FORMER WATER DISTRICTS:
71=DISTRICT 34   72=DISTRICT 42   73=DISTRICT 42   76=DISTRICT 48
77=DISTRICT 29   78=DISTRICT 29   79=DISTRICT 16   80=DISTRICT 23

COLUMN (PM) - CODE OF THE PRINCIPAL MERIDIANS REQUIRED FOR LOCATION.
S SIXTH PRINCIPAL MERIDIAN   U UTE PRINCIPAL MERIDIAN
N NEW MEXICO PRINCIPAL MERIDIAN   C COSTILLA SURVEY

COLUMN (TWN) OR (RNG) - AN (H) FOLLOWING EITHER A TOWNSHIP TWN OR RANGE RNG NUMBER INDICATES A HALF TOWNSHIP OR RANGE. AND A (U) INDICATES A UTE SECTION.

COLUMN (160) - THE 160 ACRE QUARTER SECTION

COLUMN (40) - THE 40 ACRE QUARTER-QUARTER SECTION

COLUMN (10) - THE 10 ACRE QUARTER-QUARTER-QUARTER SECTION

COLUMN (2) - THE TYPE OF USE IS CODED AS FOLLOWS:
I IRRIGATION   C COMMERCIAL   R RECREATION   F FIRE   S STOCK
M MUNICIPAL   N INDUSTRIAL   P FISHERY   D DOMESTIC   O OTHER
(***=COMBINATION OF FOUR OR MORE OF THE PRECEDING)

IF AN "A" APPEARS IN THE COLUMN JUST BEFORE, AND NEXT TO, THE USE, THIS PARTICULAR ACTION IS PART OF A PLAN FOR AUGMENTATION.

COLUMN (AMOUNT) - THE UNITS ARE CODED: CFS CUBIC FEET PER SECOND.   AF ACRE FEET

COLUMN (3) - THIS COLUMN CONTAINS A CODE DESCRIBING THE TYPE OF ADJUDICATION
O ORIGINAL   S SUPPLEMENTAL   C CONDITIONAL   CA CONDITIONAL TO ABSOLUTE
TF TRANSFER FROM   TT TRANSFER TO   AP ALTERNATE POINT   AB ABANDONMENT

| NAME OF STRUCTURE | TYP (1) | NAME OF SOURCE | WD | LOCATION 1 4 6 0 0 0 | SEC | TWN | RNG | P M | USE (2) | AMOUNT | TYP ADJ (3) | ADJ DATE | PREV ADJ DATE | APPRO DATE | BASIN RANK |
|---|---|---|---|---|---|---|---|---|---|---|---|---|---|---|---|
| JAMES MARION YOAST RES | R | W BR FISH CREEK | 57 | NWNE | 29 | 4-N | 87-W | S | I | 301.27AF | S | 06/29/1915 | 10/06/1914 | 07/31/1909 | 574 |
| MARION YOAST OUTLET D | D | W BR FISH CREEK | 57 | SESW | 20 | 4-N | 87-W | S | I | 3.3300CFS | S | 06/29/1915 | 10/06/1914 | 07/31/1909 | 574 |
| YOAST FEEDER DITCH | D | W BR FISH CREEK | 57 | NESE | 32 | 4-N | 87-W | S | I | 3.3300CFS | S | 06/29/1915 | 10/06/1914 | 07/31/1909 | 574 |
| EMRICH FEEDER DITCH | * | TEMPLE GULCH | 57 | SWSW | 06 | 5-N | 88-W | S | I | 13.7400CFS | S | 06/29/1915 | 10/06/1914 | 09/16/1911 | 575 |
| EMRICH OUTLET DITCH | D | EMRICH GULCH | 57 | | 31 | 6-N | 88-W | S | I | 5.7600CFS | S | 06/29/1915 | 10/06/1914 | 09/16/1911 | 575 |
| EMRICH RES | R | EMRICH GULCH | 57 | | 31 | 6-N | 88-W | S | I | 421.00AF | S | 06/29/1915 | 10/06/1914 | 09/16/1911 | 575 |
| MORIN DITCH | D | DAYTON CREEK | 44 | SWSE | 07 | 4-N | 89-W | S | I | 0.6600CFS | S | 06/21/1915 | 10/08/1914 | 05/15/1905 | 576 |
| MORIN RES | R | DAYTON CREEK | 44 | SWSE | 07 | 4-N | 89-W | S | I | 7.1300AF | S | 06/21/1915 | 10/08/1914 | 05/15/1905 | 576 |
| SMITH DITCH | D | SMITH CREEK | 58 | NESE | 22 | 8-N | 86-W | S | I | 1.6600CFS | S | 06/22/1915 | 10/09/1914 | 08/19/1888 | 577 |
| CORONA NO 1 WELL | W | GROUND WATER | 44 | NWNW | 07 | 3-N | 89-W | S | DS | 0.0160CFS | O | 12/31/1974 | | 12/31/1914 | 578 |
| HOMESTEAD GULCH WELL | W | GROUND WATER | 44 | SWSW | 35 | 5-N | 90-W | S | DS | 0.0670CFS | O | 12/31/1974 | | 12/31/1914 | 578 |
| MATHEWS WELL | W | AUSTRIAN CREEK | 57 | SWNW | 34 | 4-N | 86-W | S | DS | 0.0890CFS | O | 12/31/1972 | | 06/01/1915 | 579 |
| DUNKLEY DEUBEAU RES | R | WILLOW CR | 44 | SWNW | 01 | 3-N | 88-W | S | I | 112.90AF | S | 07/06/1915 | 06/21/1915 | 08/04/1904 | 580 |
| WHITELEY DITCH | D | WHIPPLE CREEK | 58 | SWSE | 23 | 3-N | 85-W | S | I | 1.6600CFS | S | 06/29/1915 | 06/22/1915 | 08/01/1902 | 581 |
| DITON DITCH | D | DRY FK TROUT CREEK | 57 | SENE | 05 | 5-N | 85-W | S | I | 0.6700CFS | S,TT | 06/19/1916 | 06/29/1915 | 05/30/1905 | 582 |
| DITON DITCH NO 1 | D | DRY FK TROUT CREEK | 57 | SENE | 05 | 5-N | 85-W | S | I | 0.6700CFS | S | 06/19/1916 | 06/29/1915 | 05/30/1905 | 582 |
| DITON DITCH NO 1 | D | DRY FK TROUT CREEK | 57 | SENE | 05 | 5-N | 85-W | S | I | 0.6700CFS | S,TF | 06/19/1916 | 06/29/1915 | 05/30/1905 | 582 |
| OSBORN DITCH | D | NORTH HUNT CREEK | 58 | NWNE | 23 | 3-N | 85-W | S | I | 0.8000CFS | S | 07/08/1915 | 06/29/1915 | 07/31/1907 | 583 |
| OSBORNE RES | R | RASPBERRY CREEK | 58 | NWNE | 23 | 3-N | 85-W | S | I | 15.5000AF | S | 07/08/1915 | 06/29/1915 | 07/31/1907 | 583 |
| DITON DITCH | D | DRY FK TROUT CREEK | 57 | SENE | 05 | 5-N | 85-W | S | I | 0.1700CFS | S | 06/19/1916 | 06/29/1915 | 07/07/1914 | 584 |
| SOUTH SIDE OUTLET D | D | DRY FORK SPRING | 57 | | | 5-N | 85-W | S | I | 0.0900CFS | S | 06/19/1916 | 06/29/1915 | 07/27/1914 | 585 |
| PICKERILL WELL | W | GROUNDWATER | 58 | SESE | 25 | 5-N | 85-W | S | D | 0.0220CFS | O | 12/31/1978 | | 07/01/1915 | 586 |
| SULLIVAN FEEDERS 1 A 2 | D | CEDAR CREEK | 44 | | 35 | 4-N | 90-W | S | I | | S | 08/18/1915 | 07/06/1915 | 06/30/1902 | 587 |
| SULLIVAN RES | R | CEDAR CREEK | 44 | SENW | 35 | 4-N | 90-W | S | I | 16.1800AF | S | 08/18/1915 | 07/06/1915 | 06/30/1902 | 587 |
| SULLIVAN SUPPLY D | D | CEDAR CREEK | 44 | SENW | 35 | 4-N | 90-W | S | I | | S | 08/18/1915 | 07/06/1915 | 06/30/1902 | 587 |
| BURNT MESA D E | D | S BR HUNT CREEK | 58 | SENW | 14 | 2-N | 86-W | S | I | 1.8300CFS | S | 07/23/1915 | 07/08/1915 | 07/15/1901 | 588 |
| BURNT MESA D E | D | S BR HUNT CREEK | 58 | SENW | 14 | 2-N | 86-W | S | I | 1.5000CFS | S | 07/23/1915 | 07/08/1915 | 07/09/1906 | 589 |
| BURNT MESA RES | R | S BR HUNT CREEK | 58 | SWNW | 23 | 2-N | 86-W | S | I | 87.2200AF | S | 07/23/1915 | 07/08/1915 | 07/09/1906 | 589 |
| BURNT MESA DITCH E | D | S BR HUNT CREEK | 58 | SENW | 14 | 2-N | 86-W | S | I | 0.3300CFS | S | 07/23/1915 | 07/08/1915 | 11/02/1911 | 590 |
| CAMPBELL DITCH | D | ELK RIVER | 58 | SWSE | 08 | 8-N | 85-W | S | I | 1.5000CFS | S,TT | 01/05/1916 | 07/23/1915 | 07/01/1895 | 591 |
| WINKELMAN DITCH | D | ELK RIVER | 58 | SENW | 08 | 8-N | 85-W | S | I | 1.5000CFS | S | 01/05/1916 | 07/23/1915 | 07/01/1895 | 591 |
| WINKELMAN DITCH | D | ELK RIVER | 58 | SENW | 08 | 8-N | 85-W | S | I | 1.5000CFS | S,TF | 01/05/1916 | 07/23/1915 | 07/01/1895 | 591 |
| WOLFE DITCH | D | LITTLE BEAR CREEK | 44 | | | | | | I | 0.9500CFS | S | 08/30/1915 | 08/18/1915 | 06/18/1902 | 592 |
| BEAVER DITCH | D | LITTLE BEAR CREEK | 44 | NWNW | 31 | 9-N | 89-W | S | I | 0.3300CFS | S | 08/30/1915 | 08/18/1915 | 08/10/1903 | 593 |
| HIGHLINE DITCH 1ST ENL | D | E FK WILLIAMS FK | 44 | NWNW | 16 | 4-N | 89-W | S | I | 4.9200CFS | S | 08/30/1915 | 08/18/1915 | 04/01/1905 | 594 |
| READER DITCH | D | POOSE CREEK | 44 | NESE | 03 | 2-N | 88-W | S | I | 2.2500CFS | S | 02/05/1916 | 08/30/1915 | 06/04/1900 | 595 |
| READER DITCH | D | POOSE CREEK | 44 | NW | 14 | 2-N | 88-W | S | I | | S,AP | 02/05/1916 | 08/30/1915 | 06/04/1900 | 595 |
| READER DITCH | D | POOSE CREEK | 44 | NESE | 03 | 2-N | 88-W | S | I | | S,AP | 02/05/1916 | 08/30/1915 | 06/04/1900 | 595 |
| DE CORA DITCH | D | YANK CREEK | 58 | SWNE | 35 | 7-N | 86-W | S | I | 0.1700CFS | S | 01/18/1916 | 01/05/1916 | 05/31/1901 | 596 |
| MORIN DITCH 1ST E | D | ELK RIVER | 58 | SWNE | 32 | 7-N | 85-W | S | I | 5.0000CFS | S | 02/03/1916 | 01/18/1916 | 04/24/1888 | 597 |
| RASPBERRY CR DITCH | D | RASPBERRY CREEK | 58 | NESW | 11 | 3-N | 85-W | S | I | 1.6600CFS | S | 02/05/1916 | 02/03/1916 | 12/24/1888 | 598 |
| LAWSON CR DITCH | D | LAWSON CREEK | 58 | SWNE | 34 | 3-N | 85-W | S | I | 1.5300CFS | S | 02/10/1916 | 02/05/1916 | 06/20/1890 | 599 |
| EGRY MESA DITCH | D | E FK WILLIAMS FK | 44 | NW | 13 | 2-N | 88-W | S | I | 3.0000CFS | S | 02/07/1916 | 02/05/1916 | 07/10/1904 | 600 |
| EGRY MESA D. 1ST ENL. | D | E FK WILLIAMS FK | 44 | NW | 13 | 2-N | 88-W | S | I | 1.3300CFS | S | 02/07/1916 | 02/05/1916 | 12/28/1904 | 601 |
| BERT BUTLER DITCH | D | BUTLER CREEK | 44 | NESW | 28 | 4-N | 89-W | S | I | 1.5000CFS | S | 06/23/1916 | 02/07/1916 | 05/01/1900 | 602 |
| ED BUTLER FEEDER D | D | BUTLER CREEK | 44 | SENW | 28 | 4-N | 89-W | S | I | 0.5000CFS | S | 06/23/1916 | 02/07/1916 | 05/01/1900 | 602 |
| SAGEBRUSH RES NO 1 | R | BUTLER CREEK | 44 | SWNW | 33 | 4-N | 89-W | S | I | 9.1100AF | S | 06/23/1916 | 02/07/1916 | 07/24/1904 | 603 |
| SAGEBRUSH RES NO 2 | R | BUTLER CREEK | 44 | NESW | 28 | 4-N | 89-W | S | I | 6.1700AF | S | 06/23/1916 | 02/07/1916 | 07/24/1904 | 603 |
| SAGEBRUSH DITCH NO 1 | D | BUTLER CREEK | 44 | SENW | 33 | 4-N | 89-W | S | I | | S | 06/23/1916 | 02/07/1916 | 10/01/1904 | 604 |
| SAGEBRUSH DITCH NO 2 | D | BUTLER CREEK | 44 | NENW | 33 | 4-N | 89-W | S | I | | S | 06/23/1916 | 02/07/1916 | 10/01/1904 | 604 |

COLORADO

| | |
|---|---|
| CONCERNING THE ABANDONMENT LIST OF)<br>WATER RIGHTS AND CONDITIONAL WATER)<br>RIGHTS IN WATER DIVISION 6 ) | ABANDONMENT LIST |

Notice is hereby given that pursuant to Section 37-92-402, C.R.S. (1973 & 1983 supp.) the Division Engineer of Water Division No. 6, in consultation with the State Engineer, has made such revisions as determined to be necessary or advisable to the July 1, 1978, abandonment list. The abandonment list contains those water rights or conditional water rights that the Division Engineer has determined to have been abandoned in whole or in part. The abandonment list when concluded by judgment and decree, shall be conclusive as to water rights determined to have been abandoned.

The July 1, 1978 abandonment list for Water Division No. 6 contained no water rights or conditional water rights that the Division Engineer determined to have been abandoned, in whole or in part. Accordingly, no abandonment list was either printed or published in 1978. The revisions to the 1978 abandonment list have resulted in a determination that certain absolute and conditional water rights have been abandoned, in whole or in part.

The list of absolute and conditional water rights determined to have been abandoned, in whole or in part, may be inspected after July 1, 1984, in the offices of the Division Engineer, each Water Commissioner and each County Clerk and Recorder at any time during regular office hours. The Division Engineer will furnish or mail a copy of the Water Division abandonment list to any requesting the same upon payment of a fee of five dollars ($5).

Any person who wishes to protest the inclusion of any water rights on the abandonment list and any revisions thereto shall file a written protest with the Water Clerk and the Division Engineer in accordance with the procedures of Section 37-92-401(5), C.R.S. (1983 supp.) except that such protest must be filed with the Water Clerk no later than December 31, 1984. The fee for filing such a protest with the Water Clerk shall be twenty dollars 20).

Dated this 1st day of July 1984.

Wesley E. Signs, Division Engineer
Jeris A. Danielson, State Engineer

WATER DIVISION 6 DISTRICT 44 DIVISION ENGINEER ABANDONMENT LIST JULY 1, 1984

| STRUCTURE NAME | DIV-ID-SEQNO | SOURCE NAME | DECREED AMOUNT | ABANDONED AMOUNT | REMAINING AMOUNT | ADJ TYPE | ADJ DATE | P ADJ DATE | APPRO DATE |
|---|---|---|---|---|---|---|---|---|---|
| TIPTON IRR DITCH | 64400785009 | DRY FK LIT BEAR CR | 5.0800 | 2.1300 | 2.9500 | 2 3 | 09/18/1905 | 10/01/1904 | 06/17/1903 |
| TIPTON IRR DITCH | 64400785002 | DRY FK LIT BEAR CR | 5.2500 | 3.8300 | 1.4200 | 2 3 | 09/18/1905 | 10/01/1904 | 01/02/1905 |
| SULLIVAN DITCH NO 2 | 64400774002 | CEDAR CREEK | 1.4200 | 1.4200 | 0.0000 | 2 3 | 09/20/1905 | 09/30/1904 | 09/20/1905 |
| BIGGS DITCH | 64400556001 | WADDLE CREEK | 0.6700 | 0.0900 | 0.5800 | 2 3 | 09/21/1905 | 09/20/1905 | 02/18/1905 |
| YELLOW JACKET DITCH NO 1 | 64400518003 | BEAVER CR T MILK CR | 2.0000 | 2.0000 | 0.0000 | 2 3 | 09/20/1906 | 06/09/1906 | 12/15/1891 |
| DILLABAUGH RES | 64403685002 | DOWDEN GULCH | 11.8000 | 11.8000 | 0.0000 | 2 3 | 06/04/1907 | 09/20/1906 | 05/25/1903 |
| ROUND BOTTOM D NO 2 | 64400750001 | YAMPA RIVER | 3.6600 | 3.6600 | 0.0000 | 2 3 | 06/22/1909 | 09/16/1907 | 03/01/1889 |
| MURPHY DITCH NO 2 | 64400718001 | DRY FK ELKHEAD CR | 2.6600 | 0.5200 | 2.1400 | 2 | 06/22/1909 | 10/08/1908 | 08/10/1902 |
| JAKODOWSKY DITCH | 64400663001 | SUGAR CREEK | 1.5000 | 1.5000 | 0.0000 | 2 3 | 06/22/1909 | 10/08/1908 | 07/15/1906 |
| HAUGHEY IRR DITCH | 64400647001 | FORTIFICATION CREEK | 2.7500 | 1.5000 | 1.2500 | 2 3 | 06/22/1909 | 10/08/1908 | 05/01/1907 |
| KNOWLES IRR D | 64400679001 | KNOWLES CREEK | 1.3300 | 1.3300 | 0.0000 | 2 3 | 06/08/1910 | 06/06/1910 | 07/16/1907 |
| DIAMOND DITCH | 64400595002 | ELK HEAD CREEK | 1.1700 | 1.1700 | 0.0000 | 2 3 | 09/23/1911 | 08/12/1911 | 11/15/1903 |
| MULLEN DITCH | 64400716001 | DEER CREEK | 1.3300 | 1.3300 | 0.0000 | 2 3 | 08/20/1912 | 08/19/1912 | 05/25/1904 |
| BAKER COTTONWOOD D | 64400542001 | LIT COT T FORTIFI | 1.3000 | 1.3000 | 0.0000 | 2 3 | 03/25/1935 | 05/31/1934 | 04/21/1902 |
| MC DONALD DITCH | 64400698004 | LIT COT T FORTIFI | 0.2500 | 0.2500 | 0.0000 | 2 3 | 04/29/1937 | 10/07/1935 | 05/10/1907 |
| LITTLE COTTONWOOD CK RES | 64403696002 | LIT COT T FORTIFI | 433.6300 | 433.6300 | 0.0000 | 2 3 | 04/29/1937 | 10/07/1935 | 07/14/1909 |
| LEFTWICH RES | 64403695002 | BOONE CREEK | 28.4000 | 28.4000 | 0.0000 | 2 3 | 12/13/1948 | 04/27/1942 | 05/15/1947 |
| COVE LAKE RES | 64403682001 | MORAPAS CREEK | 28.5000 | 28.5000 | 0.0000 | 2 3 | 09/01/1960 | 05/16/1949 | 06/01/1915 |
| COVE RES | 64403683001 | MORAPAS CREEK | 53.3000 | 53.3000 | 0.0000 | 2 3 | 09/01/1960 | 05/16/1949 | 06/01/1915 |
| ELK TRAIL DITCH | 64400611002 | PINE CREEK | 8.4000 | 8.4000 | 0.0000 | 2 3 | 09/01/1960 | 05/16/1949 | 06/07/1915 |
| ROBY DITCH NO 2 | 64400748002 | MORAPAS CREEK | 1.4000 | 1.4000 | 0.0000 | 2 3 | 09/01/1960 | 05/16/1949 | 07/28/1921 |
| UTE TRAIL DITCH | 64400789001 | MORAPAS CREEK | 2.3500 | 2.3500 | 0.0000 | 2 3 | 09/01/1960 | 05/16/1949 | 07/28/1921 |
| ROBY D AKA ROBY D NO 1 | 64400747001 | MORAPAS CREEK | 3.4000 | 3.4000 | 0.0000 | 2 3 | 09/01/1960 | 05/16/1949 | 07/28/1921 |
| NICHOLS DITCH NO 2 | 64400721002 | YAMPA RIVER | 1.0000 | 1.0000 | 0.0000 | 2 3 | 09/01/1960 | 05/16/1949 | 05/04/1945 |
| CATARACT DITCH | 64400573001 | LITTLE BEAR CREEK | 11.5000 | 11.5000 | 0.0000 | 2 3 | 09/01/1960 | 05/16/1949 | 09/13/1948 |
| MOCK DITCH | 64400711001 | YAMPA RIVER | 8.9000 | 8.9000 | 0.0000 | 2 3 | 09/01/1960 | 05/16/1949 | 08/24/1950 |
| BUFFAMS DITCH | 64400561002 | YAMPA RIVER | 4.7000 | 4.7000 | 0.0000 | 2 3 | 09/01/1960 | 05/16/1949 | 06/01/1952 |
| TIPTON IRR DITCH | 64400785006 | DRY FK LIT BEAR CR | 9.7400 | 8.7400 | 1.0000 | 2 3 | 09/01/1960 | 05/16/1949 | 10/09/1952 |
| ALBERT HORTON RES | 64404319001 | DRY FK LIT BEAR CR | 381.2000 | 381.2000 | 0.0000 | 2 3 | 09/01/1960 | 05/16/1949 | 04/15/1955 |
| D D FERGUSON D NO 2 | 64400587002 | MILK CREEK | 22.0000 | 22.0000 | 0.0000 | 2 3 | 09/01/1960 | 05/16/1949 | 05/28/1955 |
| MARTIN CK DITCH | 64400692002 | MARTIN CREEK | 24.0000 | 24.0000 | 0.0000 | 2 3 | 09/01/1960 | 05/16/1949 | 05/28/1955 |
| D D AND E RES | 64403689001 | MARTIN CREEK | 536.4400 | 536.4400 | 0.0000 | 2 3 | 09/01/1960 | 05/16/1949 | 05/28/1955 |
| MARTIN CK DITCH | 64400692001 | MARTIN CREEK | 14.4000 | 14.4000 | 0.0000 | 2 3 | 09/01/1960 | 05/16/1949 | 09/30/1958 |
| MILK CK DITCH | 64400706003 | MILK CREEK | 11.4700 | 11.4700 | 0.0000 | 2 3 | 09/01/1960 | 05/16/1949 | 09/30/1959 |
| BAKER COTTONWOOD D | 64400542004 | LIT COT T FORTIFI | 3.0000 | 3.0000 | 0.0000 | 2 3 | 05/30/1972 | 09/01/1960 | 05/01/1938 |
| FREEMAN D + ALT HGT | 64400618003 | MILK CREEK | | 0.0000 | | 237 | 05/30/1972 | 09/01/1960 | 08/03/1949 |
| FREEMAN D + ALT HGT | 64400618005 | MILK CREEK | 0.5000 | 0.5000 | 0.0000 | 2 | 05/30/1972 | 09/01/1960 | 08/03/1949 |
| BUTTS DITCH | 64400798002 | MORGAN GULCH CR | 1.7800 | 1.7800 | 0.0000 | 2 3 | 05/30/1972 | 09/01/1960 | 05/01/1950 |
| PITNEY RESERVOIR | 64403720001 | CORRAL GULCH | 11.2300 | 11.2300 | 0.0000 | 2 3 | 05/30/1972 | 09/01/1960 | 06/06/1954 |
| UTE TRAIL DITCH | 64400789003 | MORAPAS CREEK | 3.0000 | 3.0000 | 0.0000 | 2 3 | 05/30/1972 | 09/01/1960 | 09/01/1954 |
| WILSON ILES DITCH | 64402028001 | MILK CREEK | 1.0000 | 1.0000 | 0.0000 | 2 3 | 05/30/1972 | 09/01/1960 | 09/15/1960 |
| SCHNEIDERHEINZ PUMP | 64400835002 | YAMPA RIVER | 1.0000 | 1.0000 | 0.0000 | 2 3 | 05/30/1972 | 09/01/1960 | 06/29/1964 |
| WEAVER PUMP | 64400873001 | YAMPA RIVER | 0.1500 | 0.1500 | 0.0000 | 2 3 | 05/30/1972 | 09/01/1960 | 05/20/1966 |
| WEAVER WELL | 64405001001 | GROUND WATER | 0.4400 | 0.4400 | 0.0000 | 2 3 | 05/30/1972 | 09/01/1960 | 02/28/1967 |
| FIVE PINES PUMP STATION | 64400808001 | YAMPA RIVER | 2.0000 | 2.0000 | 0.0000 | 2 3 | 05/30/1972 | 09/01/1960 | 06/01/1967 |
| DUZIK PUMP | 64400805002 | YAMPA RIVER | 2.9000 | 2.9000 | 0.0000 | 2 3 | 05/30/1972 | 09/01/1960 | 05/01/1969 |
| GUIRE COLLECTION PL | 64400811001 | FORTIFICATION CREEK | 1.0000 | 1.0000 | 0.0000 | 2 3 | 05/30/1972 | 09/01/1960 | 02/21/1970 |

ADJ TYPE 1-ORIGINAL 2-SUPPLEMENTAL 3-CONDITIONAL 4-TRANSFER TO 5-TRANSFER FROM 6-ABANDONMENT 7-ALT POINT 8-COND TO ABSOLUTE

COLORADO

39

Ditch Warrior Meas. Device Cement

Stream Bear and Turkey

Priorities No. 4. 8. 14. and 16

| Date | Amt. of River Water in Ditch | Amt. of Reser. Water in Ditch | Date | Amt. of River Water in Ditch | Amt. of Reser. Water in Ditch | Date | Amt. of River Water in Ditch | Amt. of Reser. Water in Ditch |
|---|---|---|---|---|---|---|---|---|
| Nov. 1 | 8.0 | | May 1 | 12 | | June 1 | 6.0 | |
| " 2 | 8.0 | | " 2 | 12 | | " 2 | 6.0 | |
| " 3 | 8.0 | | " 3 | 12 | | " 3 | 5.0 | |
| " 4 | 8.0 | | " 4 | 12 | | " 4 | 5.0 | |
| Apr. 6 | 12 | | " 5 | 12 | | " 5 | 5.0 | |
| " 7 | 12 | | " 6 | 12 | | " 6 | 5.0 | |
| " 8 | 12 | | " 7 | 12 | | " 7 | 4.0 | |
| " 9 | 12 | | " 8 | 12 | | | 0 | |
| " 10 | 12 | | " 9 | 12 | | " 27 | 6.0 | |
| " 11 | 12 | | " 10 | 12 | | " 28 | 6.0 | |
| " 12 | 12 | | " 11 | 12 | | | | |
| " 13 | 12 | | " 12 | 12 | | | | |
| " 14 | 12 | | " 13 | 12 | | | | |
| " 15 | 12 | | " 14 | 12 | | | | |
| " 16 | 12 | | " 15 | 12 | | | | |
| " 17 | 12 | | " 16 | 12 | | | | |
| " 18 | 12 | | " 17 | 12 | | | | |
| " 19 | 12 | | " 18 | 12 | | | | |
| " 20 | 12 | | " 19 | 12 | | | | |
| " 21 | 6.0 | | " 20 | 12 | | | | |
| " 22 | 6.0 | | " 21 | 12 | | | | |
| " 23 | 6.0 | | " 22 | 12 | | | | |
| " 24 | 6.0 | | " 23 | 12 | | | | |
| " 25 | 6.0 | | " 24 | 12 | | | | |
| " 26 | 8.0 | | " 25 | 12 | | | | |
| " 27 | 0 | | " 26 | 10 | | | | |
| " 28 | 0 | | " 27 | 6.0 | | | | |
| " 29 | 0 | | " 28 | 6.0 | | | | |
| " 30 | 12 | | " 29 | 6.0 | | | | |
| | | | " 30 | 12 | | | | |
| | | | " 31 | 8.0 | | | | |

First day water was used Nov. 1 - 1953, Last day water used, Oct. 31

No. days water carried 161, Av. daily amount carried 10.1

No. acre feet used 3252, No. acres irrigated 1755

Ditch Warrior Meas. Device

Stream

Priorities

| Date | Amt. of River Water in Ditch | Amt. of Reser. Water in Ditch | Date | Amt. of River Water in Ditch | Amt. of Reser. Water in Ditch | Date | Amt. of River Water in Ditch | Amt. of Reser. Water in Ditch |
|---|---|---|---|---|---|---|---|---|
| July | | | Aug. 1 | 6.0 | | Sept 1 | 0 | |
| " 17 | 6.0 | | " 2 | 6.0 | | " 2 | 0 | |
| " 18 | 12 | | " 3 | 4.0 | | " 3 | 6.0 | |
| " 19 | 12 | | " 4 | 4.0 | | " 4 | 6.0 | |
| " 20 | 12 | | " 5 | 4.0 | | " 5 | 8.0 | |
| " 21 | 12 | | " 6 | 0 | | " 6 | 8.0 | |
| " 22 | 12 | | " 7 | 30. | | " 7 | 8.0 | |
| " 23 | 18 | | " 8 | 15 | | " 8 | 8.0 | |
| " 24 | 12 | | " 9 | 12 | | " 9 | 8.0 | |
| " 25 | 12 | | " 10 | 12 | | " 10 | 8.0 | |
| " 26 | 12 | | " 11 | 12 | | " 11 | 6.0 | |
| " 27 | 12 | | " 12 | 12 | | " 12 | 6.0 | |
| " 28 | 12 | | " 13 | 15 | | " 13 | 0 | |
| " 29 | 12 | | " 14 | 15 | | " 14 | 0 | |
| " 30 | 12 | | " 15 | 12 | | " 15 | 0 | |
| " 31 | 4.0 | | " 16 | 12 | | " 16 | 0 | |
| | | | " 17 | 12 | | " 17 | 0 | |
| | | | " 18 | 12 | | " 18 | 0 | |
| | | | " 19 | 12 | | " 19 | 0 | |
| | | | " 20 | 12 | | " 20 | 8.0 | |
| | | | " 21 | 12 | | " 21 | 8.0 | |
| | | | " 22 | 10 | | " 22 | 8.0 | |
| | | | " 23 | 10 | | " 23 | 12 | |
| | | | " 24 | 10 | | " 24 | 12 | |
| | | | " 25 | 10 | | " 25 | 12 | |
| | | | " 26 | 10 | | " 26 | 12 | |
| | | | " 27 | 8.0 | | " 27 | 12 | |
| | | | " 28 | 6.0 | | " 28 | 12 | |
| | | | " 29 | 6.0 | | " 29 | 12 | |
| | | | " 30 | 0 | | " 30 | 8.0 | |
| | | | " 31 | 0 | | | | |

Last day water used

Av. daily amount carried

No. acres irrigated

First day water was used

No. days water carried

No. acre feet used



Ditch Warrior Meas. Device

Stream

Priorities

| Date | Amt. of River Water in Ditch | Amt. of Reser. Water in Ditch | Date | Amt. of River Water in Ditch | Amt. of Reser. Water in Ditch | Date | Amt. of River Water in Ditch | Amt. of Reser. Water in Ditch |
|---|---|---|---|---|---|---|---|---|
| Oct 1 | 12 | | | | | | | |
| " 2 | 10 | | | | | | | |
| " 3 | 10 | | | | | | | |
| " 4 | 10 | | | | | | | |
| " 5 | 10 | | | | | | | |
| " 6 | 10 | | | | | | | |
| " 7 | 12 | | | | | | | |
| " 8 | 12 | | | | | | | |
| " 9 | 12 | | | | | | | |
| " 10 | 12 | | | | | | | |
| " 11 | 12 | | | | | | | |
| " 12 | 12 | | | | | | | |
| " 13 | 11 | | | | | | | |
| " 14 | 10 | | | | | | | |
| " 15 | 10 | | | | | | | |
| " 16 | 10 | | | | | | | |
| " 17 | 10 | | | | | | | |
| " 18 | 10 | | | | | | | |
| " 19 | 10 | | | | | | | |
| " 20 | 10 | | | | | | | |
| " 21 | 10 | | | | | | | |
| " 22 | 10 | | | | | | | |
| " 23 | 8.0 | | | | | | | |
| " 24 | 8.0 | | | | | | | |
| " 25 | 8.0 | | | | | | | |
| " 26 | 8.0 | | | | | | | |
| " 27 | 8.0 | | | | | | | |
| " 28 | 8.0 | | | | | | | |
| " 29 | 8.0 | | | | | | | |
| " 30 | 8.0 | | | | | | | |
| " 31 | 8.0 | | | | | | | |

Last day water used

Av. daily amount carried

No. acres irrigated

First day water was used

No. days water carried

No. acre feet used

# FEDERAL, STATE, AND LOCAL ROLES

Our multilayered system of government has a variety of divided and shared responsibilities. While a detailed discussion of intergovernmental relationships may not be appropriate here, it will be useful to summarize the basic allocation of power.

## General Allocation of Power over Water

Our system of government is organized in a unique way. The federal government has limited powers which are enumerated in the Constitution, as well as those incidental powers necessary and proper to carry out the enumerated powers. All other powers are reserved to the states. The exercise of these reserved powers is determined by an examination of each state's constitution. Local governments are political subdivisions of the state and have only those powers delegated by the state government through constitutional provisions or statutes, commonly referred to as enabling acts. The basic control and regulation of water (as is true for most issues involving property), at least with respect to intrastate matters, is left to the discretion of each individual state.

## The Federal Government

The federal government's involvement in water matters involves the exercise of two powers: the power (1) to own property and to regulate with respect to that property (making it one of the world's most powerful landowners) and (2) to regulate matters involving or affecting interstate commerce.

Regarding its own property, the federal lands, the federal government acts in two broad ways. It may exercise a reserved right and it may deny or condition permission to enter on its lands to create water diversion and/or transportation facilities.

### *Permission to Use Federal Lands*

The federal government owns a substantial amount of the well-watered land in the West and exercises substantial control over the use of that water. The owner of a state water right who diverts or stores water on federal lands or transports water across federal lands must obtain permission from the United States to be on its land. The government will condition its permission on the water user's making significant concessions, including such things as making releases to maintain or enhance stream flows or developing additional wildlife habitat. This practice has spawned significant controversy. Some describe the government's requirements as long overdue responsible resource management. Others call it blackmail.

## *Exercise of the Commerce Clause*

The federal government has the exclusive power to regulate interstate commerce. The exercise of that power has taken a variety of forms, from the navigation servitude to water pollution control.

## *Federal Reserved Rights*

Since the early 1960s, a tremendous controversy has developed over the exercise of *federal reserved rights,* formerly thought to apply only to Indian reservations. Since the term represents a concept developed solely by judicial decision, it is necessary to examine the case law regarding reserved rights to develop an adequate definition. There are innumerable reported cases on reserved rights, resulting in a myriad of contradictory decisions. Therefore, one can only rely on the decisions of the United States Supreme Court.

The case most frequently cited as the genesis of the reserved right doctrine is *United States v. Rio Grande Dam & Irrigation Co.* (1899). The United States sought to restrain the defendant from constructing a dam across the Rio Grande in the territory of New Mexico and thereby appropriating waters for irrigation and other purposes. Regarding whether a state is empowered to change the common-law riparian system of water rights and adopt the appropriation doctrine, the Court said:

> Although this power of changing the common-law rule as to streams within its dominion undoubtedly belongs in each state, yet two limitations must be recognized: *First, that, in the absence of specific authority from congress, a state cannot, by its legislation, destroy the right of the United States, as the owner of lands bordering on a stream, to the continued flow of its waters, so far, at least, as may be necessary for the beneficial uses of the government property.* (emphasis added)

That the United States possessed rights to the continued flow of water on its land was thus established.

The actual form of this right of continued flow, at least as it applies to federal reservations, was elucidated in *Winters v. United States* (1908), the first case to address the issue of reserved rights directly. The reservation in question in *Winters* was a Montana Indian reservation established in 1888. In 1889, without complying with Montana law, the United States diverted 1000 miners' inches of water from a stream bordering the reservation for use on the reservation. When later appropriations impaired the government's diversion, the United States sued to restrain them. The Supreme Court upheld the restraining order on the theory that the United States had intended, in 1888, to reserve not only land but also water, the use of which could not be impaired by the subsequently appropriating defendants. The Court said:

> The power of the government to reserve the waters and exempt them from appropriation under the state laws is not denied, and could not be. . . . That the

> government did reserve them we have decided, and for a use which would be necessarily continued through years.

The Court found that the purpose of the reservation, the transformation of the Indians from a nomadic to a pastoral people, could not be achieved without such a reservation of water. The United States was awarded 1000 miners' inches in this first case dealing directly with the reserved right doctrine.

The next significant United States Supreme Court case to address the reserved rights issue was *Arizona v. California* (1963). In that decision, the Supreme Court clearly expanded the doctrine of reserved rights to include non-Indian federal withdrawals and reservations, while also reaffirming the doctrine's applicability to Indian reservations. Regarding reserved rights claimed on behalf of the five Indian reservations in question, the Supreme Court concluded that:

1. The United States had, as a matter of fact and law, intended to reserve waters as well as lands;
2. The United States had reserved an amount of water sufficient to irrigate all the practicably irrigable acreage on the reservations. The Court specifically approved the rational of *Winters* in finding that water was necessary to the establishment of civilized communities on the reservation—the main objective of the reservation system.

The Supreme Court also extended the reservation doctrine to other classes of federal reservations. The Court stated:

> . . . the principle underlying the reservation of water rights for Indian Reservations was equally applicable to other federal establishments such as National Recreation Areas and National Forests. . . . The United States intended to reserve water sufficient for the future requirements of the Lake Mead National Recreation Area, the Havasu Lake National Wildlife Refuge, the Imperial National Wildlife Refuge, and the Gila National Forest.

The amount of water available to each of these non-Indian reservations was more specifically established in the decree of the Court wherein each of the reservations was awarded a sufficient quantity of water as was reasonably necessary to fulfill its purposes.

The next occasion for a discussion of the reserved right doctrine was the Court's decisions in the companion opinions of *United States v. District Court in and for County of Eagle, Colorado* (1971) and *United States v. District Court in and for Water Division No. 5, Colorado* (1971) (hereinafter *Eagle County* case). Both opinions involved the United States' claims for reserved water rights on the various non-Indian federal reservations in the State of Colorado. The *jurisdiction* of the state courts of Colorado to adjudicate the reserved rights of the United States was questioned. The Supreme Court found that, under 43 U.S.C. § 666, the McCarran

Amendment, the Colorado courts did have the jurisdiction to adjudicate the government's claims.

Despite the procedural focus of the two opinions, the Supreme Court did make several statements which clarified non-Indian reserved rights. Regarding the power of the United States to reserve waters, the Court in *Eagle County* noted:

> It is clear from our cases that the United States often has reserved water rights based on withdrawals from the public domain. As we said in *Arizona v. California*, the Federal Government had the authority both before and after a State is admitted into the Union "to reserve waters for the use and benefit of federally reserved lands." The federally reserved lands include any federal enclave.

As in the case of reserved rights on Indian reservations, the Court indicated that waters may be reserved to achieve the objectives of a non-Indian reservation: "The reservation of waters may be only implied and the amount will reflect the nature of the federal enclave."

In *Cappaert v. United States* (1976), the Supreme Court dealt with the reserved rights of the United States appurtenant to the Devil's Hole National Monument. The Cappaerts were the owners of a large ranch bordering the monument. In the operation of the ranch, the Cappaerts, subsequent to the establishment of the monument in 1952, began withdrawals of underground water for use on their ranch. The Cappaerts' withdrawals caused the level of water in the Devil's Hole National Monument to drop below the level necessary to ensure the survival of the Devil's Hole pupfish, a species found only in Devil's Hole and which the monument had been created to protect. The United States successfully sought to enjoin the diversions by the Cappaerts to the extent necessary to maintain the pupfish. The Supreme Court affirmed the grant of that injunction.

The Court reaffirmed the reserved right doctrine and assured its applicability to all forms of federal reservations when it stated:

> This Court has long held that when the Federal Government withdraws its land from the public domain and reserves it for a federal purpose, the Government, by implication, reserves appurtenant water then unappropriated to the extent needed to accomplish the purpose of the reservation. In so doing the United States acquires a reserved right in unappropriated water which vests on the date of the reservation and is superior to the rights of future appropriators. Reservation of water rights is empowered by the Commerce Clause, Art. I., § 8, which permits federal regulation of navigable streams, and the Property Clause, Art. IV, § 3, which permits federal regulation of federal lands. The doctrine applies to Indian reservations and other federal enclaves, encompassing water rights in navigable and nonnavigable streams.

The Court noted that water will be considered reserved only when the United States intended to do so, and that such intent may be implied if previously unappropriated waters are necessary to accomplish the purposes of the reservation. In *Cappaert*, the Court found that the reservation was express. The Court did caution, however, that only such waters as are necessary to fulfill the purposes of a particular reservation could be reserved. In *Cappaert,* it was found that water sufficient to assure survival of the pupfish had been reserved when the monument was created in 1952. Therefore, it upheld the injunction of the Cappaerts' diversions, but only to the extent necessary to maintain the level of water required for pupfish survival.

After the *Cappaert* opinion was announced, environmentalists expected that the reserved right doctrine would result in unnumbered benefits in western streams where water rights for environmental purposes were often unavailable under state law. The euphoria was short-lived, for just one year later, in 1977, the Supreme Court delivered its opinion in *United States v. New Mexico* (1977), popularly known as the Mimbres Case, in which reserved rights for national forests existed only to preserve timber and to secure favorable conditions of water flows for the benefit of downstream appropriators. The Court not only rejected national forest reserved rights for aesthetic, recreational, wildlife preservation, and stockwatering purposes, but also emphasized that the reserved water right exists only for the *primary* purpose of any land reservation; water rights for secondary uses must be acquired under state law.

The previous cases permit the extrapolation of a broad definition of the federal reserved right doctrine. It can be authoritatively stated that:

1. The United States has the power to reserve certain unappropriated waters appurtenant to any federal reservation or withdrawal from the public domain.
2. Whether the United States actually intended to exercise that power by reserving water in any such instance must be determined in connection with specific claims of the United States.
3. The intent of the United States to reserve water may be implied by the actions and circumstances surrounding a particular reservation.
4. Any such reserved right exists only to serve the purposes of the reservation and later may be utilized only on the reservation to effectuate the purposes for which it was created.
5. The reserved right appurtenant to any federal reservation is for that quantity of water which is reasonably necessary to fulfill the purposes of the reservation.
6. Within an appropriation system, the date of priority of a particular reserved water right is the date of the reservation.

## The State Government

The state governments have the power to define property interests, establish how those interests are created and conveyed, and regulate those property interests (including water rights) through the exercise of their police powers. Consequently, the most important level of government with respect to water matters is state government. There are, however, some limitations on state authority. Not surprisingly, those limitations are the powers of the federal government. For almost a century, the courts have struggled with the integration of federal reserved rights with state priority systems. In addition, states are proscribed from acting in such a way as to substantially affect matters of national or interstate interests. State efforts to prohibit the export of water beyond their own boundaries have been overridden as impermissible burdens on interstate commerce—an area over which the federal government has plenary control.

### Local Agencies

When water matters are best dealt with on a local basis, the states have not hesitated (in fact may have gone overboard) in allowing the formation of local entities to deal with specific water problems. In addition to municipalities, for which water is big business, there are various quasimunicipal corporations or special districts which exercise substantial power and influence over water. These range from local, neighborhood water districts which distribute water among paying consumers, to groundwater management districts which have the power to regulate the use of groundwater within their borders. Each such district, being a creature of state statute, is a special entity. The statutes and organic documents of each state and each district must be examined carefully to determine the powers and limitations attributable to that district.

## WATER ORGANIZATIONS

The early history of the western United States is one of remarkable individual effort and sacrifice. The earliest water rights were appropriated by individuals who obtained land close to a water source, making construction of an individual ditch possible even in the absence of mechanized assistance. Innovation and cooperation became necessary as settlement of the West progressed. The land taken by subsequent settlers was often far from a water source, which made it difficult, if not impossible, for individuals to develop their own water rights and their own ditch systems. A cooperative effort was required. The history of these cooperative efforts may be one of the finest hours in the development of the West.

### Ditch Companies

The first cooperative efforts were two or three neighbors getting together to dig a ditch and create a water right which they shared. As long as the number of irrigators was kept to a minimum, the joint use of a ditch was a perfectly acceptable solution to the need to pool resources.

However, the opportunities for limited joint efforts were quite few. Usually, a large number of landowners got together, pooled their money, and formed a *nonprofit* corporation, which would construct a large ditch or canal or reservoir to supply water to its members. The annual operational and maintenance costs for the project were allocated among the members in the form of assessments. Unfortunately, the boom and bust syndrome, so familiar in the West, created a sense of instability for these *mutual* ditch companies. While liens for unpaid assessments could be placed on the members' shares, foreclosures of those liens were so difficult and distasteful that they were loath to be made.

As eastern wealth came west, its holders realized an opportunity to make money in the water business. Since groups of settlers were often too poor to gather enough capital to construct even joint water facilities, entrepreneurs decided to handle that task for them. The entrepreneurs typically funded a corporation themselves, constructed facilities, and perfected water rights. The corporate owners held title to the water rights and the facilities but would enter into contracts with individual water users to supply water to them each year. Those corporations, which were *for profit,* came to be known as carrier ditch companies. Frequently, they were treated as a special type of public utility with contract rates being set by a governmental entity, often the county in which they were located.

## Districts

Governmental entities got involved when private capital no longer was adequate to fund the size of common projects needed to support development in the West. The first such entities, usually referred to as irrigation districts, raised money through revenue bonds which were paid off by taxes on *irrigated* lands. In other words, those persons who received the direct benefit of the project paid for the project. Unfortunately, many other people benefited indirectly from the project. Because irrigation provided greater income to farmers and ranchers, they spread more money among the local merchants, who paid no taxes to support the project. Consequently, a new type of district, a conservancy district, was established and enjoyed a tax base that included *all taxable property* within its boundaries.

## Municipalities

Perhaps the most significant players in the western water game today are municipalities. Recognizing that their growth is directly related to the availability of water, they have shown remarkable diligence in developing their water supplies. In the absence of any significant coordination, it has been every municipality for itself. As a result, municipalities are primarily responsible for setting the market value of water rights and for the significant retirement of irrigation water rights. One of the major issues involving municipal water efforts has been the extent to which they may monopolize the water supply for areas outside their boundaries. Recognizing that the control of new water supplies ensures the control of surrounding growth, many municipalities have thoughtfully concluded that they should be the sole provider of water in suburban areas. This attitude and effort has created controversy and conflict. The extent to which strong municipalities will supplant private enterprise or other local governmental entities in the development of western water resources is perhaps the major issue facing the water community today.

# 4

# GROUNDWATER AND WELLS

All water above and below the earth's surface is an integral part of the hydrologic cycle. Water that penetrates the earth's surface and infiltrates the soil is called subsurface water. Subsurface water may be pulled back to the surface by capillary attraction and evaporated, thus short-circuiting a part of the hydrologic cycle, or it may be absorbed by plant roots and returned to the atmosphere by transpiration. Groundwater is that which percolates deep enough to reach the zone of saturation—the reservoir that supplies water to wells. Groundwater is replenished by infiltration of precipitation and surface water. All of the water in the groundwater reservoir is not available for use because of limitations imposed by quality, dependability, access, and cost.

The principal difference between groundwater and surface water is that surface water is visible and its occurrence and movement can be observed and comprehended. Groundwater, on the other hand, cannot be seen and its occurrence and movement have historically been the subject of conjecture and misunderstanding. As early as 1843, an English court referred to groundwater as "flowing through the hidden veins of the earth" and concluded that "no man can tell what changes these underground sources have undergone in the progress of time" (*Acton v. Blundell*). This situation has led to the development of legal and administrative concepts based on the assumption that groundwater was physically separate and distinct from surface water. Such assumptions are not based on facts and have resulted in widespread misunderstanding of the occurrence and behavior of groundwater.

One of the most ancient and exhaustively debated aspects of groundwater is that of dowsing or water witching, the practice of locating underground water by use of a forked stick or other form of divining rod. An investigation of the dowsing phenomenon was published by the United States Geological Survey in 1917, and concluded:

> It is doubtful whether so much investigation and discussion have been bestowed on any other subject with such absolute lack of positive results. . . further tests by the United States Geological Survey of this so-called 'witching' for water, oil, or other minerals would be a misuse of public funds. (*U.S.G.S. Water Supply Paper 416*)

More important to the development of groundwater resources is the lack of understanding of the physical concepts of the occurrence and movement of groundwater. This perpetuates the problems of integrating groundwater and surface water rights and administration. The various legal doctrines relating to groundwater have been discussed in an earlier chapter. The following discussion is intended to present the physical concepts of groundwater as an aid in reconciling the facts with the various legal doctrines. Figure 13 illustrates the general relationship among physical, technical, and legal groundwater systems.

## OCCURRENCE AND MOVEMENT OF GROUNDWATER

Groundwater geologists will admit—if pressed—that they are no better at observing underground conditions than other people. They can, however, by application of scientific methods, obtain an understanding of the conditions that occur underground and influence the occurrence and movement of groundwater. Groundwater is primarily a function of geology and an understanding of geology and hydrology is essential to an understanding of groundwater occurrence and behavior.

A considerable amount of research has been completed to establish and verify the physical processes which govern the movement of underground water. These principles have been derived through analysis and interpretation of data from wells and springs, including well logs, monitoring of water level fluctuations, chemical analysis, discharge data, aquifer testing, and other methods. A clearer understanding of the water resources of a region can be gained by analysis of the relationship of groundwater to the water in other phases of the hydrologic cycle, both under natural conditions and man-made conditions.

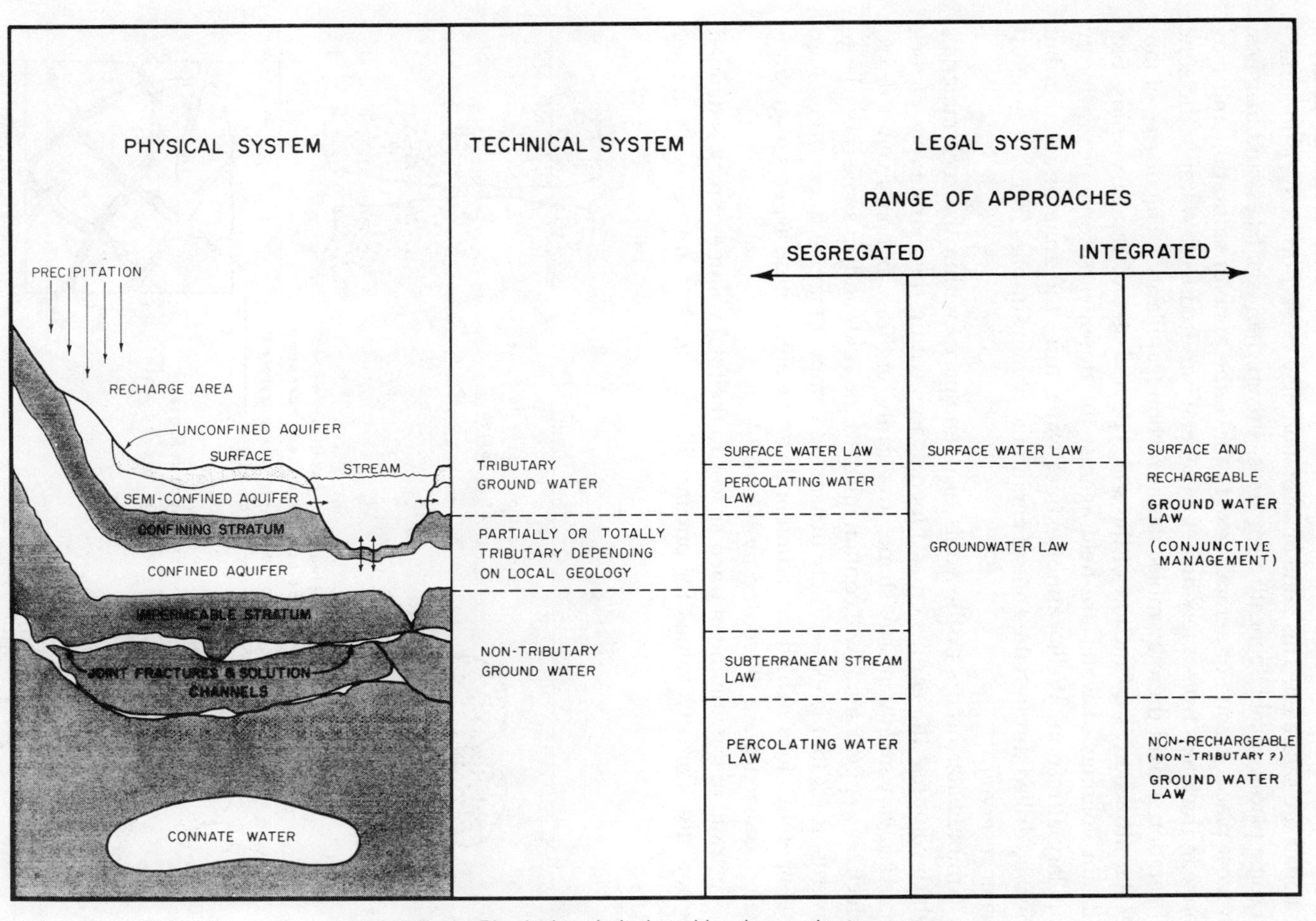


**FIGURE 13.** Physical, technical, and legal groundwater systems.

Not all water beneath the land surface is groundwater. Groundwater is that part of the subterranean water that occurs where all voids in the containing materials are saturated. This zone of saturation may extend up to the land surface in some places, notably beneath seep areas, stream channels, lakes, and marshes. In most places a zone of aeration exists above the zone of saturation and may range in thickness from a few inches to several hundred feet. Water in the zone of aeration is held there by molecular attraction (capillarity). The degree of molecular attraction is a function of the soil type and size of capillary voids. Significant volumes of water are held against the downward force of gravity by capillary attraction. Wells cannot extract water from the zone of aeration but must be drilled through the zone of aeration to obtain supplies from the groundwater reservoir.

Groundwater occurs in the voids or open spaces within the rock materials which comprise the earth's crust. Open spaces within the many types of rock materials vary greatly in size, shape, irregularity, and distribution. Even bedrock, which may appear solid, will contain microscopic voids and, in some cases, large caverns. Rock that is exceedingly dense and compact may be fractured enough that openings form which can admit and store water. These aspects of groundwater occurrence are illustrated by Figure 14.

The ratio of the volume of open spaces to the total volume of rock is defined as porosity. Porosity values for unconsolidated deposits are generally higher

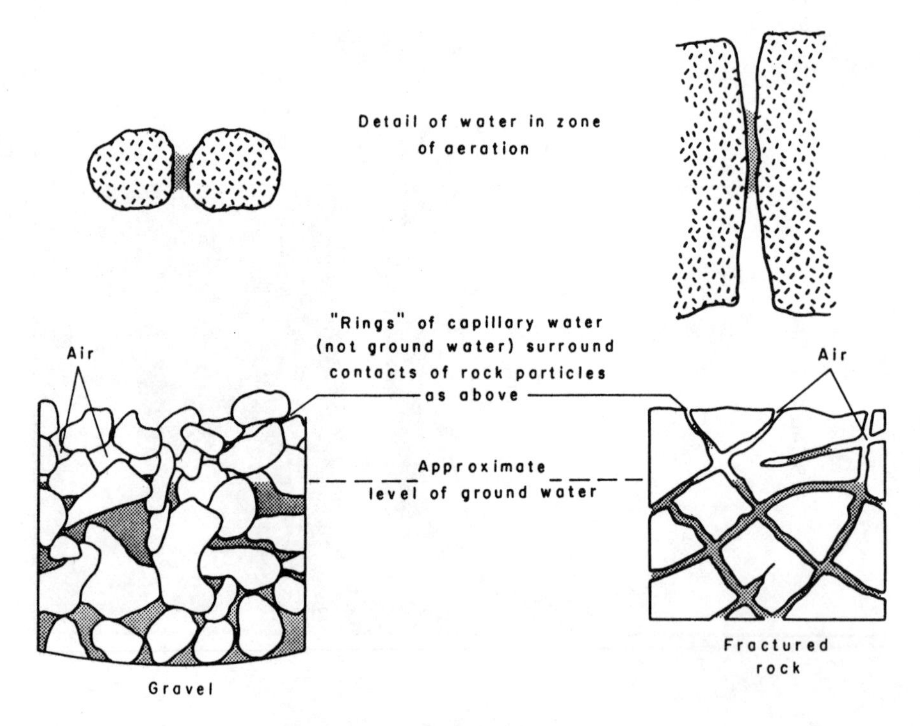


**FIGURE 14.** Water occurrence in rocks.

than for consolidated rock. Porosity is not simply a function of the void size as evidenced by the fact that fine sand and clay deposits have higher porosities than coarse sand and gravel deposits. The porosity of unconsolidated materials will vary with the packing, shape, and sorting of the particles. Grain size is not a determining factor, because if other conditions are the same, a material will have the same porosity whether it consists of small or large grains. In some consolidated rocks the original porosity has been reduced by compaction or by the formation of cement in the pore spaces. In other cases, it may have increased by development of fractures or by the dissolving of the rock material.

Rocks of low porosity are limited in their capacity to absorb, hold, or yield water. Such rocks may occur close to the earths surface, although on the average, the porosity tends to be higher near the surface and lower as depth increases. The least porous rocks are those that are buried so deeply that the weight upon them distorts the rock structure and closes all pores.

Rock materials of high porosity may yield large quantities of water, but not necessarily. One saturated rock may yield most of the water in its pores to wells or springs while another of equal porosity but smaller pores may retain practically all its water and yield negligible amounts to wells. The difference is in the degree of hydraulic conductivity or interconnection of the voids.

As the size of pore space declines, molecular attraction becomes more significant because of the larger surface area of solid material to which water can adhere. A cubic foot of well-sorted quarter-inch gravel will have a combined surface area of the grains of about 200 square feet, while a cubic foot of well-sorted sediment of rounded, clay-size grains, 0.001 millimeter in diameter, with a porosity equal to that of the gravel will have a surface area of the grains of more than a million square feet. The molecular attraction for water in the clay will be correspondingly greater. Rock that allows water to move by gravity is more permeable than rock that holds water by molecular attraction.

Groundwater movement is primarily a function of the permeability of the rock material, which is its capacity of transmitting water. In permeable materials, gravity moves water downward from the land surface to the zone where all pores are saturated, then laterally toward lower elevations, and ultimately to the oceans or places where the water is discharged at the land surface by springs or seeps, or consumed by evaporation or transpiration.

## DEVELOPMENT OF GROUNDWATER

Development of groundwater by wells involves drilling or digging a hole into the zone of saturation. Water draining from the saturated rock materials into the well is called gravity groundwater. As that water is pumped out, other water moves toward the well. The rate at which water moves toward the well, and therefore the rate at which water can be withdrawn from the well, depends largely on the transmissivity of the materials from which the water is drawn.

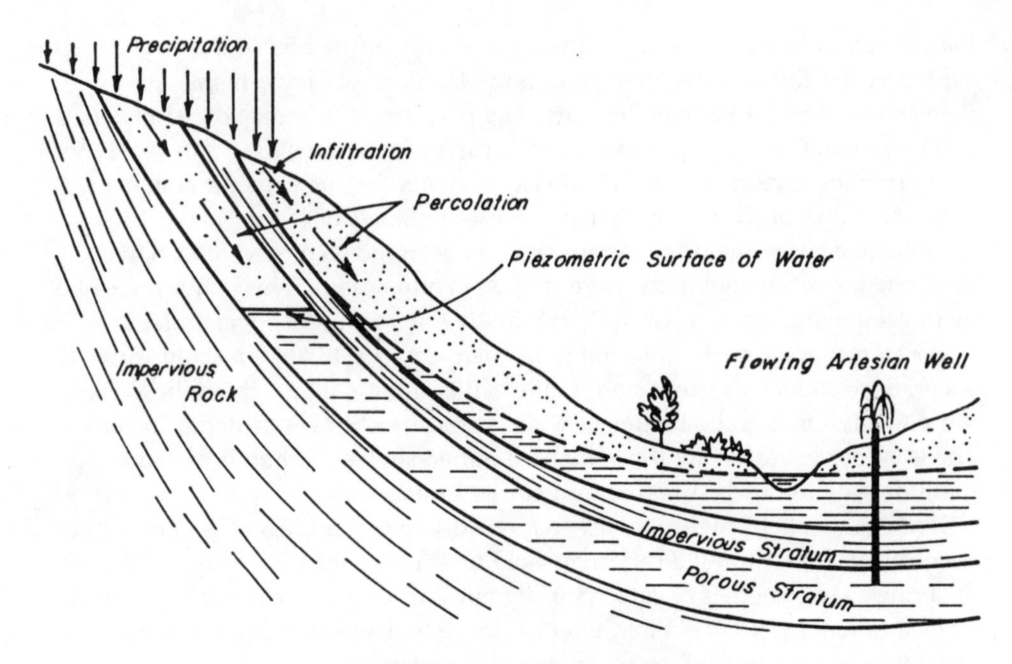


**FIGURE 15.** Artesian groundwater.

Waterbearing geologic formations or deposits that yield usable amounts of groundwater are referred to as aquifers. The term aquifer only indicates an ability to yield groundwater and has no quantitative limitations. A water supply of less than one gallon per minute supplying one household may be an adequate aquifer, whereas aquifer yields of thousands of gallons per minute may be required for supplying municipal and industrial water systems.

Groundwater aquifers generally occur under confined (or artesian) or unconfined conditions. A confined aquifer is separated from the atmosphere by materials that restrict the movement of water into or out of the aquifer (Figure 15). In a confined aquifer, the water is under pressure and will rise in a well bore, or deep hole, to a level (a potentiometric surface) above the bottom of the overlying confining layer. In some cases, the pressures are great enough to cause water to flow at the surface. When this occurs, the well is referred to as a flowing artesian well. groundwater, within unconfined aquifers, is in direct interconnection with atmospheric pressures.

The word *artesian* comes from the town of Artois in France, the old Roman city of Artesium. It was at Artois that the best known flowing artesian wells were developed in the Middle Ages.*

Deep wells that reach far below the water table are often called artesian wells, but this is an incorrect use of the term. Such deep wells may withdraw unconfined water. The word artesian can be properly used only when the water is under

* The role of Artesians in artesian wells is not well understood, but is thought to be connected with Murphy's law in some way.

pressure, and the aquifer is confined beneath or between layers of impermeable rock.

When groundwater is not confined under pressure, it is described as occurring under water-table conditions. For practical reasons it is necessary to know whether groundwater is under artesian or water-table conditions. It can be assumed that artesian water is continuous for some distance beneath the confining layer of rock, and that it is replenished by water which enters the aquifer some distance away. Under water-table conditions, groundwater is recharged locally and is more immediately responsive to precipitation. An artesian supply is less reliable for long, continued use than an unconfined supply, which is recharged more locally.

The term groundwater reservoir is often used interchangeably with aquifer. In a broad sense, it may represent the waterbearing materials in an extensive area, for example, the San Joaquin Valley groundwater reservoir in California, which supplies water equal to about a quarter of all the water pumped from wells in the United States. The term has been used more loosely to designate all the rock material of a continent, but such broad usage ignores the local variations in porosity and the resulting variations in capacities to store and yield groundwater.

Variations in porosity and permeability of rock materials are responsible for the variation in occurrence of water in the groundwater reservoirs of the continent. In some areas, the aquifers and the materials above it are all relatively permeable, although some variation is almost inevitable. Wells dug or drilled in those areas usually penetrate some dry strata, then materials that are moist (the capillary fringe), and then materials from which water enters the well; thus they have gone through the zone of aeration and reached the zone of saturation. The water-table rises as recharge enters an unconfined aquifer; it falls as water drains away or is discharged by natural means or wells.

Relatively impermeable layers within the zone of aeration produce complications in this simple picture. Such layers may be right at the surface, so that there can be very little infiltration, and practically all the water from precipitation collects on the surface of the permeable layer or runs off overland.

Impermeable material may be present at various depths below the land surface, whether as hardpan within the soil zone, or clay or other tight material at greater depth. Such material may be poorly permeated and thus retard downward percolation, so that there is an accumulation of water and temporary saturation of material immediately above it. However, it may be so impermeable as to prevent downward percolation, so that a groundwater reservoir is permanently maintained above the impermeable zone. Such bodies of groundwater within the zone of aeration are perched, as illustrated in Figure 16. Many are of very small extent and may disappear in a short time after rains. Some perched groundwater reservoirs yield water to wells and springs and can be of economic value. Their reliability as a supply source is dependent on the size of the permeable layer and amount of recharge.

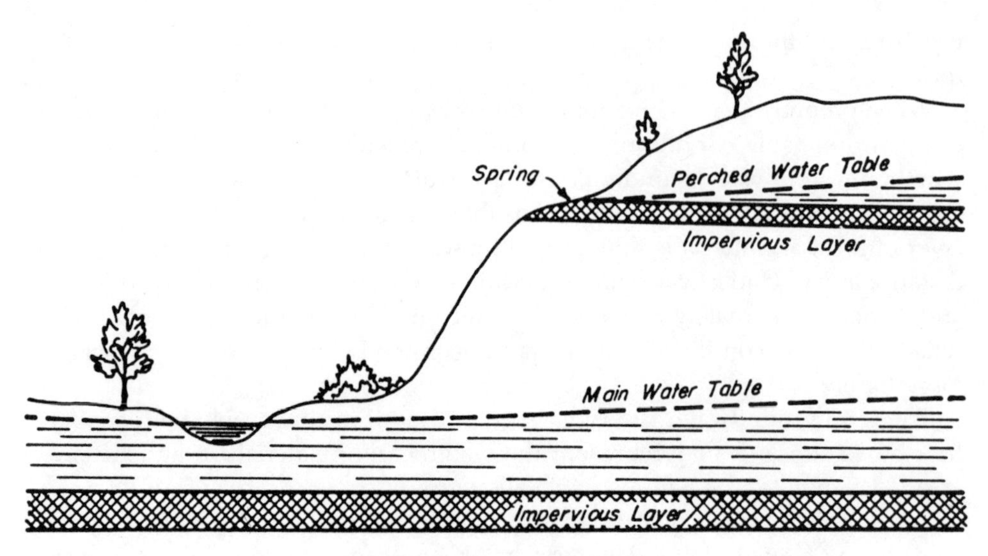


**FIGURE 16.** Perched water table.

Many deep wells penetrate far into the zone of saturation and go through rock materials that vary widely in their permeability. Water may rise markedly above the top of the aquifer and may even overflow at the surface. Such water is confined beneath less permeable material. It is not necessary that the confining bed be impermeable, but only that it be less permeable than the aquifer. The Dakota sandstone in North Dakota and South Dakota is a famous artesian aquifer in which water is confined beneath dense shale. At the base of the Wasatch Range in Utah, there are beds of loose sand more permeable than the Dakota sandstone; nevertheless, that sand constitutes the confining bed above coarse gravel. In both instances water moves far more easily through the aquifer tapped by wells than through the overlying beds.

Artesian aquifers are not recharged by downward movement through the overlying confining beds. The artesian pressure prevents downward movement where the confining bed is moderately permeable. Artesian aquifers are generally recharged in areas where confining beds do not exist and where groundwater is under water-table conditions. It is important to differentiate between confined and unconfined waters because of the differences in groundwater hydraulics. Artesian aquifers are only part of the underground system necessary for developing supplies to wells or springs. For artesian and other aquifers, there must be some area in which the soil or other surficial material, and any underlying unsaturated material, are sufficiently permeable to permit water to recharge the aquifer.

The volume of storage in an aquifer does not correspond to the capability of the aquifer to provide a sustained yield of water to wells and springs. The limit of sustained yield is governed by the average annual recharge to the aquifer, just as the useful yield of a surface reservoir is governed by the inflow to it. Alluvial aquifers adjacent to streams are often capable of producing large sustained yields because they are readily recharged by the stream. Others, notably in desert areas, may be larger and hold large quantities of water in storage, but yet have very low capabilities for sustained yield because of limited recharge.

Water which penetrates the earth's surface and becomes groundwater, behaves in some ways similarly to surface water. One major difference is that groundwater moves more slowly than surface water. Surface water flow is described in feet per second, whereas groundwater flow is thought of in feet per year. There is a greater amount of time for groundwater to dissolve the material through which it moves, but less velocity to scour and physically transport that material. Groundwater contains more dissolved solids, but less suspended solids, than surface water. The greater concentration of certain dissolved solids is the reason groundwater is hard and requires chemical softening before use. An important function of the hydrologic cycle is water treatment by means of evaporation during the recycling process.

Both intrinsic permeability and hydraulic conductivity are important factors in determining the sustained yield of wells. That yield is limited to the quantity of water that moves to the well from the recharge area. If wells remote from a source of recharge annually yield volumes of water greater than the annual volume of water recharged to that part of the aquifer, mining of the aquifer will occur.

Artesian supplies in particular are limited by the transmissivity of the aquifer, and many problems of failing water supplies stem from this limitation. The problems are analogous to those of a town that has an adequate overall water supply, but has distribution lines too small to carry the water needed in some parts of the town.

The safe yield of an aquifer (using the term to describe an underground unit of the hydrologic cycle, with its areas of recharge and discharge) cannot be greater than the average recharge, whether that recharge occurs by natural or artificial means, or both. The quantity recovered for use from the entire aquifer will be appreciably less than that inflow, because of the practical limitations of the recovery techniques. In any part of the aquifer—even at any particular well—a further limitation is imposed by the rate at which water can be transmitted through the aquifer. This limitation is least pertinent to wells within the recharge area, where they can reap the benefit of any inflow. The limitations are an oversimplification of a complex problem and are only the hydrologic limitations. In many areas economic, social, and legal factors limit the yield of groundwater aquifers to a rate far below their hydrologic capabilities.

An analogy exists between recharge and sustained yield of aquifers and inflow and yield of surface reservoirs. Groundwater development programs are not based on knowledge of the recharge comparable to knowledge of streamflow because developed surface storage does not exist until a substantial investment has been made in labor and materials, whereas the groundwater reservoir is developed and filled by nature and is ready to yield water for a relatively small investment in well drilling. Groundwater shortages have been reported in areas in nearly every state. Most of the difficulties could have been mitigated if the hydrology of the groundwater reservoirs had been known and understood.

In many areas the surficial materials are permeable but unsaturated. In some places such materials may extend to depths of several hundred feet. They do not come within the definition of aquifers or groundwater reservoirs, which are limited to saturated materials, but they can be classed as potential underground reservoir sites, which may be placed in service if suitable methods can be found for saturating them. Groundwater reservoirs exist in the irrigated areas of the West that once were composed of unsaturated rocks but that now receive the excess water from irrigation, store it, and release it to wells and springs. There also are unsaturated rocks that will probably never provide satisfactory reservoirs, although they are exceedingly permeable, because the materials are in places where water would drain out as fast as it could be added.

Groundwater recharge generally moves downward from the surface. The rate and volume of recharge is a function of the permeability of the overlying soil or rock and the available water from precipitation, streams, or other sources. Many recharge areas are underlain by materials so permeable that they would not be classed as soils because water moves down through them so rapidly that they support little vegetation. Examples of this are the talus on mountain slopes, the bouldery beds of streams at the mouths of canyons, barren lavas, sand dunes, and areas of outcrop of cavernous limestone.

In most aquifers, only a small part of the infiltration to the recharge area becomes groundwater. Near the land, surface water may be consumed by evaporation or transpiration and returned to the atmosphere. In deserts, most of the water from the limited precipitation is dissipated in this way, and the groundwater reservoirs may be recharged only a few times a century, during exceptional wet years.

Groundwater and surface water interact when rivers and streams flow through aquifers. When the stream is located below the adjacent water table, water will flow to the stream and the surface flow will be gaining. When the stream surface is below the water table, water will flow to the aquifer and the stream will be losing flow. These conditions are illustrated in Figure 17.

Even in humid areas, many aquifers receive negligible recharge during the summer. This indicates that vegetation may consume most, if not all, of the water made available to the aquifer by rains during the growing season.

Water percolating below the root zone is beyond the influence of the sun's energy, and its movements are impelled by gravity. It moves in response to gravity as it goes down to the zone of saturation, but it also moves in response to pressure as it rises in spring orifices or flowing wells or seeps upward through so-called confining layers. Most of the movement in the groundwater phase is lateral, down gradient, rather than vertical. It is similar to movement of surface water in this respect.

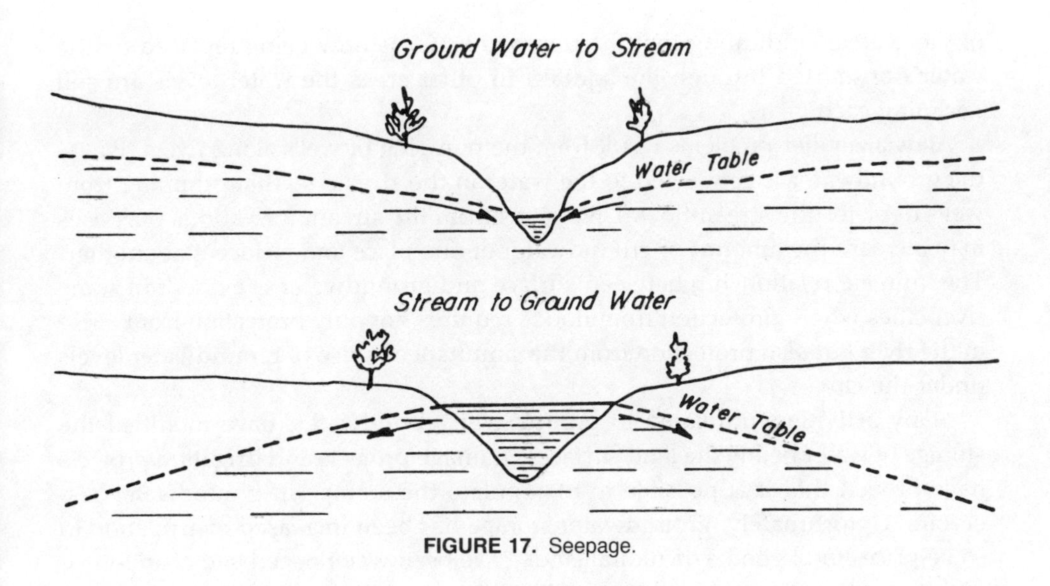


**FIGURE 17.** Seepage.

# WELLS

Several problems are created by pumping from wells. Some pertain to entire groundwater reservoirs, where the rate of replenishment is inadequate to meet the continuing demand. Pipeline problems (problems arising from inadequate capacity of aquifers to transmit needed quantities of water to points of use) arise because of the inability of water to move rapidly enough through earth materials to supply the demand of wells, even though the aquifers, as a whole, may have an adequate supply of water. Other problems occur along surface streams due to the close relation that exists between the water in the stream and that pumped from wells.

Serious problems of groundwater shortage occur in areas where water is pumped out faster than the entire groundwater reservoir is replenished. Under those conditions, the reservoir is being emptied of water that may have taken decades or centuries to accumulate, and there is no possibility of a continuous perennial supply unless existing conditions are changed. Most of the excessively pumped aquifers are in arid regions where precipitation is generally inadequate for the needs of man. Users of groundwater are generally aware that they are using more than the perennial supply and that the supply will be exhausted unless action is taken. Corrective measures already applied in some areas include prevention of waste, a pro rata reduction of pumping from all wells, prohibition of further development, reclaiming of used water, artificial groundwater replenishment by surplus stream water, and importation of water from other areas.

Pumping from closely spaced wells has caused significant local declines in water levels, chiefly in municipal and industrial areas that use large quantities of groundwater. The water levels have reached approximate equilibrium in some of those areas, indicating that the pumped water is now being replaced by the water transmitted through the aquifer. In other areas the water levels are still declining each year.

Alluvial aquifer problems result from the pumping of wells along rivers, where the groundwater is connected to the water in the stream so that pumping from wells depletes the streamflow. Diversions from the stream for various purposes may increase the amount of groundwater at one place and reduce it at another. The intimate relationship between surface and groundwater is evident in some river cities where protection from floods requires not only protection from a rise in the river but also protection from the simultaneous rise of groundwater levels under the city.

Many activities, unrelated to pumping of groundwater, have modified the storage of water below the land surface. Drainage projects and irrigation projects have proved that it is possible to manipulate the storage in groundwater reservoirs. Unfortunately, groundwater storage has been increased by irrigation in some places until good agricultural lands have been waterlogged and abandoned, and it has been decreased in other localities by drainage to the detriment of agricultural use of the land or municipal use of the water. The quantity of water stored underground has been altered by surface water storage structures and flood protection works, channel improvements and urban storm sewer systems. Water supplies have been polluted by discharging contaminated water into streams from which it enters groundwater reservoirs, and also by puncturing protective layers, thus permitting entry of sea water or other mineralized water into aquifers.

## ADMINISTRATION

The complexity of the groundwater system and resulting lack of understanding of the occurrence and movement of groundwater have historically led to the development of legal concepts and doctrines that are only vaguely related to the facts. This has resulted in the development of numerous classifications of groundwater for administrative and judicial purposes.

Factually, the most realistic classification of groundwater is the distinction between tributary and nontributary. Tributary groundwater is physically connected, or tributary, to the surface stream system. Withdrawal and use of this groundwater depletes the surface stream in a manner similar to a surface diversion. Because the wells which are used to withdraw tributary groundwater are located at a distance from the flowing stream and groundwater moves at a slower velocity than surface water, the impact of withdrawal by tributary groundwater wells will be reduced in quantity and delayed in time compared to a surface water diversion. Nontributary groundwater is not connected to the surface stream system and can be withdrawn without measureable impact on surface water rights, as shown by Figure 18.

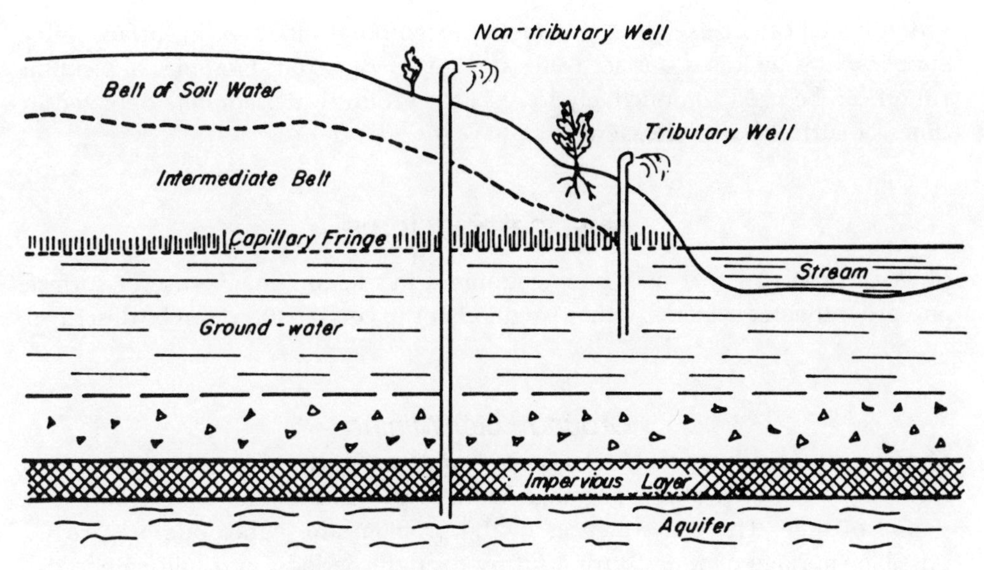


**FIGURE 18.** Groundwater zones and belts.

The question of whether groundwater is tributary or nontributary is a matter of fact and is directly related to the geology of the area concerned. The geohydrologist can, within limits, quantify the depletion to surface streams that will result from groundwater withdrawals. Both the amount and timing of depletions can be calculated using accepted analytic techniques. In Colorado, techniques have been developed specifically for evaluating the impact of pumping wells on surface streams (see, for example, Glover, 1978). Once the quantity of depletion has been determined, the results can be presented to administrative and judicial bodies for decision as to whether the water is tributary or nontributary. In Colorado, for example, the courts had significant difficulty with that issue. Eventually, groundwater that would reach a stream in less than 43 years was considered to be tributary. Subsequently, it was held that groundwater was nontributary if it would reach a stream or its pumping would affect the stream only after 100 years. The status of groundwater falling between those two extremes was left to conjecture. Finally, in 1985, the legislature stepped in to define nontributary groundwater as that which will not, within a period of 100 years, deplete the flow of a stream at an annual rate greater than one-tenth of 1 percent of the annual rate of withdrawal from the well being pumped.

# CONJUNCTIVE USE

## Definition

Conjunctive use is a management technique that takes advantage of the different characteristics of multiple sources of water. It is particularly applicable to surface and groundwater systems in which the relatively constant availability of groundwater can be used to supplement the temporal variations in surface water supplies. In some cases, surface water storage reservoirs filled in times of plentiful runoff can be used conjunctively to recharge groundwater aquifers depleted in times of surface water deficiency.

## Physical Problems

Physical problems that may be encountered in the conjunctive use of surface and groundwater systems, either singularly or in combination, can be described as follows.

### *Groundwater Mining*

Mining is generally encountered in arid and semiarid areas and occurs when recharge of groundwater is less than withdrawal from the same system over a period of time. The consequences of the problem are higher pumping costs, possible shortage of water during future drought periods, and, ultimately, depletion of the aquifer. When surface water is available in an area, the conjunctive operation of surface water and groundwater sources can reduce the impacts of the mining problem. Safe yield is the rate at which water may be extracted without mining. There is, however, water in excess of this amount which may be economically mined for brief periods. Operation under a traditional safe yield approach might incur large opportunity costs because the water resources cannot be put to optimum or near optimum use. Change from the traditional safe yield management strategy to more efficient resource use would be subject to social acceptance.

### *Saltwater Intrusion*

Saltwater intrusion is the shoreward movement of water from the ocean into confined or unconfined aquifers (usually coastal) and the subsequent displacement of freshwater from these aquifers. This is the usual problem in coastal areas or in areas adjacent to saltwater bodies where overpumping of aquifers can result in saltwater intrusion. Conjunctive operation of groundwater and surface water can mitigate the extent of intrusion. Optimum management might involve the regulation of withdrawals and recharge with respect to time and space for groundwater and temporal regulation for surface water allocation.

### *Low Flow Maintenance in a Stream Connected to an Aquifer*

In a connected aquifer–surface water system, groundwater withdrawal can influence the surface water quantity. A large withdrawal of groundwater reduces the streamflow and flow time delay effects are important. Withdrawal of groundwater usually occurs during low flow periods and may reduce the quantity of surface water, violating low flow criteria of the stream and/or impairing surface water rights. Conjunctive management of a surface water–groundwater system can reduce the impacts of groundwater withdrawal upon streamflow quantity and quality by imposing spatial and temporal schedules for groundwater withdrawal and surface water diversion.

### *Interaquifer Water Transfer*

The diversion of surface water from a river system may result in shifting water from one aquifer to another aquifer system. The result is usually a continuous decrease in one aquifer storage (and/or water level) and increase in the other aquifer storage or nearby surface water system.

### *Adverse Groundwater Quality (Surface or Surface/Groundwater Supply)*

Maintaining recommended standards for water quality usually involves water treatment. To reduce the cost of treatment facilities, conjunctive utilization of surface water and groundwater might be attempted to preserve both quality and quantity.

### *Aquifer Recharge Using Treated Wastewater*

Recharging an aquifer using treated wastewater has been proven worthwhile under special circumstances and subject to operational constraints and ultimate use of the water. The operation is useful in increasing freshwater storage, in developing a barrier against saltwater intrusion, and as a method of receiving wastewater effluent. Conjunctive analysis of surface–groundwater-treated wastewater interactions is necessary to understand the total system as well as for establishing operational policy for future groundwater extraction.

### *Aquifer Recharge Using Potable Water*

In Colorado, studies are being made to evaluate the technical and economic feasibility of recharging deep bedrock aquifers with potable water. The purpose of the studies is to demonstrate that a potable surface water supply can be injected into and recovered from a potable aquifer supply.

Along the front range metropolitan area east of the Rocky Mountains, municipal supplies are derived from both surface water and groundwater sources. Under current conditions, it is estimated that the bedrock aquifers that are used for groundwater supply will experience significant reduction in productivity as water is pumped from the potentiometric condition to a strictly confined condition.

Demand for potable water reaches a peak in the summer months of June, July, and August. During the rest of the year, there is excess capacity available in municipal water treatment facilities. This excess capacity can be used to supply treated water to the deep groundwater system to maintain the potentiometric head on the bedrock aquifers.

The studies include assessment of the chemical compatibility of the source water with the aquifer water. Legal and institutional concerns involving long-term ownership of the water and the need for adjudication are also being considered.

## Operation And Management

### *Physical Variables*

The characteristics and variation of surface water and groundwater flow are important factors in a conjunctive use operation. The dominant sources of surface water are streams, reservoirs, lakes, springs, and snow packs. Availability of quantities of water from these sources in space and time is an important factor in design and/or allocation problems. These variables are used in models of complex conjunctive use systems. The stochastic nature of these variables (especially stream flow) is important in allocation problems and should be considered in system operation. Availability, quality, and losses are three major areas of concern in the use of surface water.

Aquifer capacity and characteristics are important elements in analyzing groundwater flow where aquifer recharging systems are treated as an input to the groundwater system. Temporal and spatial variation of the recharging system outputs are also significant. In other cases, water losses from the aquifer and the quality of groundwater are important. Four principal features of groundwater recharge of aquifer systems are listed below.

1. Type and characteristics of aquifer(s) in the system. The characteristics of an aquifer include storage capacity and hydraulic properties.
2. Mechanisms for losses from aquifer(s). These include transfer of groundwater to adjacent aquifers or streams, pumping, and evapotranspiration.
3. Aquifer recharge mechanisms. Recharge mechanisms may involve precipitation (local and distant), seepage through streambeds, irrigation and irrigation return flow, transfer of groundwater either from another aquifer (horizontally or vertically) or through artificial recharge practice.
4. Quality of groundwater.

### *Data Availability*

A factor which usually dictates the level of analysis (the hierarchical scale at which the problem can be analyzed) is the availability of data. Complex conjunctive use problems require a variety of data. While many of the data may be available, different types, for instance water levels, flow, and quality, may be incompatible in form, particularly with regard to space time increments and scales. Nevertheless, available data should be utilized whenever possible to supplement well-thought-out acquisition of new data. Since the objective for managing a conjunctive use system is the optimum development of water and related land, data collection should emphasize those elements which affect water development and use both in economic and physical terms. Data required for management and operation of the system may differ from data used in analysis of the system. Availability and convenience of obtaining these data are important factors in implementing operation policy.

## *Management Objectives*

Conjunctive use management objectives include identification of current problems as well as projection of future problems. Another objective involves determination of optimum or viable solutions to the problems identified at different time horizons. While there must be an economical justification for a specific policy, other factors, such as equity among users, water quality conditions, or social and political considerations will influence the adopted solution. As a result, most problems require multi-objective analysis. Major objectives include (from highly specific to poorly defined):

1. Minimizing total water cost to satisfy a set of demands.
2. Maximizing the total net benefits generated by economic activities that use the groundwater/surface water system.
3. Maintaining acceptable water quality.
4. Achieving equity among water users.
5. Enhancing social well being.

# 5

# ENGINEERING ASPECTS

## WATER RESOURCE PROCEEDINGS: NEGOTIATIONS THROUGH LITIGATION

### Types of Proceedings

The adjudication of new water rights, the changing of existing water rights from one type, time, and place of use to another, and the resolution of conflicts between water users involving the determination of priorities or quantification of entitlement normally involves some type of administrative or judicial proceeding. Water rights matters are often included in zoning and development hearings before city, county, or regional agencies and in disputes involving real estate transactions, eminent domain proceedings, and land sales. The attorneys for the parties involved will have need for one or more experts to develop, document, and present the factual evidence relating to the case at hand.

The type of proceedings range from informal negotiations to complex litigation. Various procedures are used in compiling, documenting, and presenting the facts of a case to opposing parties or the court. These range from informal discovery by the exchange of information between experts to sworn testimony and cross-examination. If the opposing parties cannot agree on the informal exchange of information, it is often obtained through interrogatories and depositions.

Interrogatories are a list of questions prepared by one party to obtain information from the opposing party. The engineer will assist the lawyer in preparing and answering interrogatories. Very specific questions are necessary to avoid vague responses that do not provide the information needed from the interrogatories. For example, instead of one question asking how historic consumptive use was calculated, it is better to prepare a sequence of questions asking for the technique used, the source of data included, the crop coefficients adopted, the period of study, the location of climatic stations for weather and temperature data, and other significant factors. The questions prepared by the engineer are stated in the proper legal format by the lawyer for submittal to the opposing parties. It is important to answer only the specific question posed and not volunteer additional information.

Depositions consist of oral testimony, taken under oath, prior to a trial or proceeding. Questions and answers are recorded and transcribed by a court reporter. Direct and cross-examination are conducted by attorneys representing the parties involved. The rules of the deposition proceedings are similar to those of the court, although the process may be less formal. Lawyers taking depositions generally have their own experts in attendance to assist in preparing questions and evaluating responses.

All information provided in interrogatories and depositions is evaluated by opposing counsel and experts and can be used during the trial or other proceedings. In many cases the tools of discovery, such as interrogatories and depositions, are used to probe the opposition's case, to identify gaps in the evidence, weaknesses in technical approach or conclusions that can be challenged later.

## The Attorney–Expert Team

Effective management and operation of the water resource system under the prior appropriation doctrine depends on the ability of water rights engineers and experts in related disciplines, including hydrology, geology, and watershed science, to define and explain facts. The legal and administrative system utilizes the engineer's conclusions in the following manner:

1. As the basis for understanding the scientific principles and methods applied to the accumulation and categorization of knowledge in an organized and logical manner.
2. As a component in the adjudication and administrative process relying on knowledge relating to a specific factual issue.
3. To promote improvements in procedures and methods utilized in developing and managing water resources.
4. As a contribution to the establishment and implementation of water resources policies at the local, state, and national level.

A fundamental difference exists between the roles of the engineer and the lawyer. A lawyer is an advocate responsible for presenting the client's case in the most favorable manner possible. Lawyers are interested in obtaining specific facts that advance their client's interest and/or defeat the arguments of the opposing side. Engineers are interested in collecting all the data they can and studying a range of possibilities, not all of which will have a direct bearing on the case at hand.

While the engineer is not an advocate in the same sense as the lawyer, he does represent a client and is often involved in negotiating the resolution of issues for the client. The engineer provides information that has specific, strategic value which may or may not be advantageous to his client's use. All of the information developed by the engineer must be communicated to the lawyer and the client, who make the decisions as to what information may be used and what strategic course of action is adopted.

There are two important aspects of factual information. Good information is costly to acquire and use and this cost involves time, money, and manpower which are specific to the primary user. The benefits of the information are not easily controlled and, once put in use, tend to diffuse rapidly. Information can be modified and used by others, often to their advantage. It can add to the overall knowledge or confusion of other members of the proceeding. An expert's information can also be focused and controlled by the attorney arguing the case. The attorney who marshalls expensive information with strategy in mind accrues an advantage. On the other hand, an attorney who is careless in the collection and use of information can be at a disadvantage.

Equally important is the fact that all technical information does not have the same significance. Lengthy detailed accounts of technical matters are usually not effective in negotiations or trial procedures. The amount and type of scientific information included in negotiating sessions and testimony is a tactical question determined by discussion between the attorney and the engineer.

The engineer and lawyer work together to obtain the objectives of the client. The engineer is responsible for developing the facts which the lawyer must evaluate in terms of the legal issues involved. It is essential that both the engineer and lawyer know what the other expects and the role each is to play. These can be summarized as follows:

1. What the engineer expects from the lawyer:
   a. A willingness to listen and learn.
   b. A complete explanation of the problem and the issues.
   c. An understanding of the scope and limits of the engineer's field of competence.
   d. A willingness to use additional technical experts as required.
   e. Enough time and money to do the job right.
   f. A clear understanding of the engineer's scope of work and technical responsibilities.

g. An opportunity to reach conclusions based on the facts.
h. An explanation of the proceedings and rules in effect.
i. A discussion of the other side's position.

2. The attorney's role as conceived by the engineer:
    a. *Manager*—The engineer expert usually expects the attorney to assume the role of manager of the team, in addition to his traditional legal role as advocate of the client. The attorney must thoroughly understand the legal theories and issues, and be able to direct the team of experts in approaching the technical issues in a comprehensive, integrated manner and present these to the court in a way that develops and explains the factual issues of the case. The attorney may be inundated with details from the expert and must sort through this myriad of detail and identify those points which are the most essential to the case. The lawyer must meet early and continuously with the experts involved and gain at least a basic understanding of the technical issues. The lawyer must communicate a basic understanding of the legal theories involved to the experts. Ultimately, the lawyer must organize the presentation of the case, including the evidence developed by the experts, in a manner that is effective to the court.
    b. *Strategist*—The lawyer must analyze the opposition's case and anticipate the technical and legal issues to be presented. The lawyer must evaluate the opposing counsel and experts, identify strengths and weaknesses, and anticipate the line of attack to be taken by the opposition. Development of an effective strategy to counter the opposition at an appropriate time, is a necessity.
    c. *Coordinator*—The successful outcome of water resource proceedings is usually determined by a logical and carefully presented legal argument, supported by clear presentation of the evidence. The attorney must coordinate the efforts of one or more experts representing different disciplines in the development of the factual evidence. Careful attention must be given to the information developed by the experts and communicated to the attorney. At this point, the attorney must keep in mind that the expert is not an advocate, but a professional, who cannot bend the facts to suit a particular, preconceived legal theory which the lawyer wishes to advance. A frank and open development of the case will avoid a legal position unsupported by facts, or possible perjury by the expert.
    d. *Evaluator*—Because water resources cases depend on the joint performance of the attorney and the experts, the attorney must be a good listener and evaluator of the facts presented. Assimilation of the testimony of both sides is essential. Fruitful lines of direct and cross-examination, can be uncovered by thoughtful consideration of the presentations.

e. *Leader*—Many proceedings involve a number of attorneys and experts who are not always in agreement on the overall objectives and the immediate strategies to be followed. Experts can become so involved in the detail of their own analysis that they lose sight of the larger objectives of the case. It is the lawyer, serving as team leader, who must arbitrate these conflicts and decide on a course of action that will produce a favorable outcome to the client.

3. What the lawyer expects from the engineer:
   a. Technical competence.
   b. Honesty, integrity, and good judgment.
   c. The ability to articulate and defend conclusions.
   d. The presence to withstand rigorous cross-examination.
   e. Assistance in developing strategy and negotiating with other parties.
   f. Preparation of exhibits to clearly communicate the issues, facts, and conclusions.
   g. Quick, accurate response to the need for additional data and analyses.
   h. A willingness to present all the facts, even those adverse to the client's interest.
   i. Professional performance.
4. The engineer's role as conceived by the lawyer:
   a. *Investigator*—A starting point in any case is the basic research relating to the factual issues at hand, normally involving a field inspection, literature search, and compilation of readily available background materials. It is necessary to collect and organize basic data, including streamflow records, climatic measurements, diversion records, topographic maps, and other factual material.
   b. *Communicator*—Following the collection, compilation, organization, and analysis of the basic data involved, the engineer must be able to communicate in a clear, concise manner the technical fundamentals and concepts which are involved. This includes an explanation of the range and reliability of the data, the applicability of analytic techniques, including their weaknesses, the relevance of other techniques and approaches, and other factors that will be important to the attorney in developing the overall strategy.
   c. *Planner*—The expert must participate in developing the basic approach to the case. This requires an understanding of the legal issues and the ability to relate the evidence being developed to those issues. It is important for the expert to communicate to the attorney those factual matters which may be adverse to the case, as well as those factual matters which may support the argument. The expert must be able to anticipate the approach likely to be taken by the opposition, including analytic techniques, basic data, and possible conclusions. In doing this, the

expert must recognize that the attorney may have to assimilate information and advice from several experts. It is important, to present the information in a concise manner and avoid unnecessary details which are liable to be more confusing than helpful.

d. *Producer*—The expert usually has less time to do more work than he would like. In many cases, the expert is not authorized to undertake the work until late in the process because clients and attorneys tend to be optimistic about the possibility of settlement without going to trial. This sometimes places a burden on the expert and his staff in assembling and organizing the factual material in a manner that can effectively communicate the results to the attorney and the court. The expert must be able to digest and analyze the data and exhibits prepared by the opponents. An important element of this process is the maintenance of clear perspective of the case and an ability to separate matters of importance from trivial detail.

e. *Expert Witness*—The lawyers expect the experts to provide testimony, under oath, presenting the factual evidence and an opinion as to the meaning of that evidence. To satisfactorily fill this role, the expert must participate in preparation for the trial, assist the lawyer during the trial, and provide direct testimony and withstand cross-examination.

## Pretrial Preparation

Preparation for the proceeding or trial, including development of testimony, is an important step in the process. It provides the expert with an opportunity to assist the attorney in developing the case. A concise outline of the testimony is essential to ensure that the testimony will successfully communicate the important factual issues.

Trial preparation starts with the initial decision of whether to become involved in the case. This decision is based on an appraisal of the information available at the beginning, and whether the expert can willingly and ethically support the interest of the client. In water cases, this often involves identification and discussion of existing or potential conflicts of interest. It also requires an evaluation of the lawyer that will be leading the case to ensure personal and professional satisfaction with his qualifications and reputation. Representing an attorney who is poorly qualified or inadequately prepared can seriously injure the reputation of the expert witness.

It is necessary for the engineer to establish a definite understanding with the attorney regarding the need for adequate and complete investigation of the facts. This includes a discussion of the level of effort needed and the time and cost involved to satisfactorily develop the case. Once this has been established, the field and office work necessary can be undertaken. The evidence developed must be documented utilizing maps, drawings, photographs, and other items that explain the method of investigation, basis of interpretation, and conclusions.

The engineer must be aware that this supporting material will be used to explain scientific principles and empirical judgments to an audience that will include experts in the field and laymen not familiar with the technical details.

As the investigation proceeds, the attorney must be kept informed about the results obtained. This includes educating the attorney, and possibly other experts, as to the technical ramifications of the investigations and findings. Facts uncovered which may have an adverse affect on the case must immediately be brought to the lawyer's attention and their potential impact discussed. This and other information developed during the investigation, can be used during negotiations and/or in the preparation of interrogatories or the taking of depositions from opposing experts.

A general outline of the testimony should be developed. Testimony should respond to specific questions that provide relevent facts and conclusions developed and presented in a logical sequence. Extending testimony beyond that which is necessary to make the case, may be disadvantageous.

It is essential to be prepared and confident of the information to be presented before testifying. This may involve five to ten hours of preparation for each hour of testimony. Preparation should include a review of all work and reports bearing on the case, as well as relevant technical journals, professional papers, textbooks and published writings by experts in the field which may be used by opposing experts to provide questions for cross-examination. Preparation of testimony and development of the case also involves review and evaluation of depositions and interrogatories prepared by experts representing the opposition.

When giving a deposition or testifying it is permissible to refer to notes to clarify a point or recite specific data. The court and opposing counsel, however, may request and obtain copies of any notes, correspondence, reports, or other materials referred to during deposition or testimony.

The expert must furnish the attorney with a detailed professional resume, detailing background and experience in the field of expertise and pertaining to the subject of testimony. The attorney must demonstrate educational background and professional experience so the court accepts the witness as an expert in the field. This qualification is subject to cross-examination by opposing counsel who may seek to limit testimony to a narrow area or discredit the witness entirely.

The tactics of presenting evidence rely on the attorney and not the engineer. Attorneys normally have a definite opinion concerning the sequence of the presentation, which is often in a more formal and rigid framework than that of the engineer. The court procedure is designed to elicit the truth in an orderly manner and to provide a complete record for review by appeal courts.

## Testimony And Trial Assistance

Expert testimony generally starts with the introduction of the witness's credentials. This includes questions relating to education, practical experience, professional registration and licensing, publications, and other factors relating specifically to the subject of the testimony. The ultimate objective of the testimony is to provide a factual basis for an opinion rendered as an expert in the field.

The attorney will introduce the evidence through a series of questions proceeding in a logical manner to a conclusion. The final questions will be: Have you formed an opinion, based on the facts available regarding the matter in question? and What is that opinion? The expert must then clearly state that opinion and explain its basis. This is the essence of the proceeding and the most important part of the expert testimony process.

In some cases attorneys have difficulty properly phrasing questions. The expert can request that the question be rephrased or may rephrase it as part of his answer. The following principles are beneficial in presenting expert testimony:

1. Avoid technical jargon and use simple words when possible. When technical terminology is required, provide a concise definition.
2. Limit testimony to subjects within the area of expertise. Speak clearly and audibly and listen carefully to the lawyer's questions. The answer must be heard and understood by the judge, the jury, if appropriate, and the court reporter. The court reporter will appreciate having any difficult terms or names spelled for the record.
3. Ask that unclear questions be repeated or restated. If a question is too complex for a simple answer ask for it to be restated as a series of questions and clearly explain the logic and reasoning of the answer. If an answer is not known, say so.
4. Answer only the question asked and do not elaborate unnecessarily. Allow the lawyer to determine the order and presentation of evidence.
5. Admit a mistake or qualify an answer if necessary.

Cross-examination by opposing counsel clarifies testimony, exploits deficiencies in the analysis presented, identifies inconsistencies in the facts and, if possible, discredits the expert. In responding to cross-examination, it is important to be careful in answering questions and consistent in staying within the area of expertise. The following principles can be helpful in preparing for and responding to cross-examination.

1. Do not allow answers to be rushed, even though a cross-examiner tries to elicit a hasty response. Be deliberate and selective, refusing to accept confusing rapid-fire questions.
2. Do not hesitate to provide simple and obvious answers promptly.
3. Beware of compound, complex questions starting with hypothetical situations. Insist that the cross-examiner simplify the questions so that they can be answered without fear of contradiction. This includes trick questions and questions designed to elicit conflicting answers with testimony provided through depositions or interrogatories.
4. Do not hesitate to state that you have talked the case over at length with

the attorney you represent. If asked whether the attorney told you what to say, the best response is that he told you to tell the truth.

5. Explain, if asked, that you are a professional and your investigation and service is provided on a fee basis. Since the expert's opinion is necessarily objective and impartial, it is important that the fee not be contingent upon the outcome of the trial or proceeding.
6. Maintain composure and do not be goaded into intemperate or inappropriate responses. It is never wise to attempt clever responses to questions.
7. Do not underestimate the understanding of a cross-examining attorney who may pretend to be unaware of the facts and ignorant of the technical aspects of the case. Usually the attorney is as well informed on the key issues and factual matters as are the experts.
8. Do not hesitate to question statements in textbooks and professional papers by other authorities if familiar with the subject and knowledgeable of the author's background or work. Rely on published books or papers only to the extent you are familiar with the contents and confident that they are relevant to the subject of your testimony.
9. Listen carefully and remain silent when your attorney objects to a question. Be aware that your lawyer may have perceived a point that needs clarification or discussion or is hinting about a trap later to be used.
10. Do not hesitate to ask for a break and a chance to develop the information needed, if a question requires additional study or computation or reference to materials not readily available.

The expert witness advises the attorney of the strengths and weaknesses of the case as it proceeds. Offers of settlement may be made or received prior to the beginning of the actual hearing or trial. These offers must be evaluated with respect to the acceptability to the client and the possibility of gaining a better result through the hearing or trial process. The time and cost involved in prolonged litigation must be weighed against the advantage of immediate settlement.

As the trial proceeds, the expert must hear the testimony of others and assist the attorney in developing cross-examination. It is important to identify significant areas of weakness or inconsistency and develop a line of questions that will demonstrate the basic principles involved. Numerous questions on minor or unrelated points will often confuse the issues and aggravate the audience. The expert is advised to stick to the point, avoid getting entangled in trivial detail, and say as little as possible, but as much as necessary.

## SOURCES OF INFORMATION

The accurate determination of the factual situation relative to a water right issue requires the collection and analysis of basic data. One of the principal responsibilities of the water engineer is to be aware of all sources of relevent information and to keep abreast of data sources and data collection technology. Information is available from a variety of local, state, and national agencies, from water attorneys and engineers, and from water using organizations.

The type of information generally required will consist of basic topographic and geologic data to define the physical setting, aerial photographs, diversion records and agricultural statistics to determine historic use patterns and transcripts of judicial proceedings, planning studies and related materials to establish the institutional setting. Principal sources of water-related information are discussed briefly below.

*U.S. Geological Survey.* Topographic maps at a scale of 1 = 24,000 with various contour intervals covering either a 7 ½-minute or 15-minute quadrangle area, provide a standard source of topographic information. These quadrangle maps are supplemented by other topographic maps at varying scales covering state, county, and other units. The U.S. Geological Survey maintains a computerized system for storage and retrieval of water data collected by the survey in cooperation with state and local agencies. The National Water Data Storage and Retrieval System (WATSTORE) is operated and maintained on the survey's computer facilities at its national center in Reston, Virginia. The WATSTORE system consists of various files in which data are grouped and stored by common characteristics and data collection frequencies. Files are maintained for:

1. Surface water, quality of water, and groundwater data measured on a daily or continuous basis.
2. Annual peak values for streamflow stations.
3. Chemical analysis for surface and groundwater sites.
4. Water data parameters measured more frequently than daily.
5. Geologic and inventory data for groundwater sites.
6. Summary data on water use.
7. An index file of sites for which data are stored in the system.

A National Water Data Exchange (NAWDEX) is operated by the U.S. Geological Survey to assist users of water data in the identification, location, and acquisition of water data. NAWDEX encompasses four major areas of operation: identifying sources of water data, indexing of water data, data search assistance, and access to water-data bases. The U.S. Geological Survey also publishes reports of geologic and hydrologic investigations and other special studies as Water Supply Papers, Professional Papers, and other publications. The U.S. Geological Survey is also a valuable source of aerial photographs.

*U.S. Forest Service.* The U.S. Forest Service publishes maps of the national forests at various scales, showing principal cultural and topographic features. The Forest Service is another source of aerial photographs and also publishes special reports having to do with research on forest maintenance, streamflow, sedimentation, snow depth, and other specialized subjects.

*U.S. Soil Conservation Service.* The Soil Conservation Service is also a source of aerial photographs, which are used for mapping and analyzing agricultural activities including irrigated area. The Soil Conservation Service, in connection with county agencies, also undertakes cooperative programs to map surfical soil types and characteristics on a county basis.

*National Oceanic and Atmospheric Administration (NOAA).* This agency is responsible for the collection, compilation, organization, and publication of climatological data on a state basis. These data include measurements of daily, monthly, and annual temperature and precipitation with supplemental recording of evaporation, wind direction, snow cover, and other climatological data. Snow depth and water content surveys are conducted in cooperation with local agencies and used to prepare water-supply forecasts which are published on a regular basis.

*National Technical Information Service (NTIS).* National Technical Information Service is the basic source of information on government-sponsored projects and published reports. Through NTIS various abstracts and newsletters are available on a variety of subjects, including water resources and water pollution.

*State Engineer's Office.* The state engineer is responsible for the collection, tabulation, and publication of water rights data, including decreed water rights, diversions to ditches, and storage in reservoirs. The extent and type of data available varies from state to state. In Colorado, the state engineer regularly publishes a tabulation of all decreed water rights, which can be produced in a variety of formats and sorted in terms of stream, priority, and structure. The state engineer's office is also responsible for maintaining records of diversions, reservoir content, changes in decreed points of diversion and use, and other information.

Other state agencies responsible for collecting and publishing water information include water development/conservation agencies, including conservation and conservancy districts, division water courts, and state geological surveys. These agencies are generally responsible for information in connection with local districts, flood plain designation, water project planning and funding, and local geologic studies. In Colorado a particular source of information is the division water court, which publishes a monthly resume of water rights applications and can furnish copies of applications and court proceedings.

At the local level, city, county, and state planning offices and councils of governments are a source of local and regional planning data, including aerial photographs, topographic maps, flood plain mapping locations, population projections, water resource development plans, water treatment and distribution systems, and legislative data. Special districts and other agencies responsible for urban flood plain management, basin authorities, and special commissions are also a source of water related data.

Information on the historic use of water rights and the administration and operation of irrigation ditches and water systems can also be obtained by interviewing local residents, water commissioners, ditch riders, and others responsible for operation and administration.

## MODELING

A hydrologic model is a technique for imitating historic events and forecasting the future response of a hydrologic system under a variety of conditions. Models are particularly useful for evaluating the impact of a water rights action on the surface and groundwater system to identify and quantify potential injury to other vested water rights. Water rights actions, such as a new appropriation, a change in time, type, and place of use, or from direct flow to storage and a transfer of a point of diversion to a new point or alternate point of diversion, will impose new or different demands on the stream system in terms of the quantity and timing of diversions and returns. As streams become more fully appropriated and uses change from seasonal to year-round, accompanied by an increase in the value of water and degree of administration, it becomes increasingly important to accurately predict the impact of proposed changes on water rights so that equitable and administrable terms and conditions can be imposed to prevent injury.

Models, properly used and interpreted, can greatly facilitate this process by allowing the investigator to examine a larger number of alternatives to be used as a basis for negotiation of terms and conditions imposed by the court to limit injury. While the term modeling normally refers to a simulation process accomplished by computer, it can also refer to a simple operation accomplished by hand.

Models and simulation techniques are classified in several ways. Digital models incorporate discreet mathematical procedures accomplished manually or by computer to represent the situation modeled. Analog models utilize a continuous physical analogy (for instance, the flow of electricity through wires to represent the flow of water in pipes) to imitate a system response. General purpose models have been developed for application to generalized situations, such as reservoir operation, flood routing and groundwater flow. Such models are often limited in applicability to specific situations because of the assumptions that must be incorporated in their operating logic to satisfy the criteria for general applicability. Examples of this are river basin operation models that deal with diversion in an upstream to downstream direction only, and not by priority order and thus do not represent a real world situation when applied to the operation of rivers under the appropriation doctrine. These limitations can be overcome by developing models for specific systems. In many cases specific purpose models can, with some alteration of logic, be applied to other similar operations. An example of this is the application of a model developed specifically to simulate river basin and water rights administration in northwestern Wyoming, which was successfully modified for application to a Colorado basin by incorporating minor changes to reflect the difference in water rights identification and description between the two state systems.

Simulation of hydraulic events and water rights administration is accomplished by dynamic models that utilize time-variable processes which may be descriptive or conceptual. Descriptive models are of more interest to the practicing hydrologist because they incorporate observed phenomena and empirical techniques based on fundamental principles. Such models not only come closer to representing the real world, particularly with respect to water rights, but are easier to explain and interpret to attorneys, courts, and other parties to water rights actions. Conceptual models utilize theoretical principles to interpret hydrologic phenomena rather than to represent physical processes. A simplified example of the difference between descriptive and conceptual models is the use of observed runoff data for streamflow in a descriptive model, as opposed to generating streamflow by stochastic or probability techniques for use in a conceptual model.

The advantage of using computer models for water rights and hydrologic investigations are generally accepted and include the reduction of manual labor and the ability to analyze more alternatives in greater detail. If a computer model is developed for a specific situation, it imposes the need for a clear understanding of the problem and a complete definition of the system being modeled. It is also important to recognize the limits of computer simulation. These include difficulties inherent in duplicating real systems, particularly water rights administration, which does not always go by the book, but reflects the judgment and feel for the river of the administrators. Computer logic is often simplified at the cost of accuracy to accomodate these problems.

In modeling complex systems, it is sometimes necessary to simplify the system described to accommodate programming constraints. The lack of historic data describing the operation of the system modeled combined with the need to incorporate simplified procedures often makes it difficult to fully verify the accuracy of the model. This is particularly true of water rights models developed to represent strict administration when applied to a stream historically having very loose administration. It is important to avoid attributing a false sense of accuracy to the model results. The availability of a computer simulation model may encourage the generation of large volumes of output to satisfy a great number of "what if" questions. If not properly documented and presented, the results produced can be overwhelming in terms of paper, underwhelming in terms of value to the user, and can defeat the intended purpose.

Certain basic procedures can be applied to the application of simulation modeling to hydrology and water rights situations. The first step is to define the system to be modeled in terms of the location and interrelationships of the basic components, including rivers, ditches, reservoirs, exchanges, transbasin imports and exports, and other major features. This is accompanied by the compilation of a data base and the establishment of criteria for system operation. The system components should be displayed graphically by maps and diagrams to document the location and identification of inflows and outflows, diversions, returns, storage, and other operational features.

The data base is comprised of surface runoff records, diversion records, description of decreed water rights, tabulation of irrigated acreage and crops grown, and similar materials providing a factual basis for defining the system. System operation criteria include study periods to be used, administrative procedures, level of detail selected for reservoir operation, and related factors.

Construction of the model involves selecting the necessary technique, in terms of mathematical equations and accounting procedures, processing of the data base to extend incomplete records through hydrologic correlation, verification of data for accuracy and consistency, and development of the operating logic. As the model is developed, each component should be tested and calibrated to the extent possible to provide the basis for model accuracy. This is an iterative process that continues as the model is used to improve the acceptability and dependability of the result.

Figure 19 illustrates the general content and process of the Intergrated River Operation System (IROS) model, originally developed and used in the adjudication of water rights, including federal reserved rights and Indian claims in the Bighorn Basin of northwest Wyoming, and subsequently adopted to the Yampa River Basin of Colorado for use in evaluating the impact of federal instream flow claims for Dinosaur National Monument, located at the extreme downstream end of the Yampa River Basin. This is an example of a digital descriptive model developed for a specific project and later modified for general application to model river basin operation and water rights administration under the appropriation doctrine.

## DETERMINATION OF HISTORIC CONSUMPTIVE USE

Historic beneficial consumptive use is the measure of a water right and its determination is important in the appraisal of the value of water rights and in establishing the basis for developing terms and conditions imposed on water right changes to prevent injury to other vested water rights. Consumptive use is defined as diversions less returns, the difference being the amount of water physically removed (depleted) from the stream system through evapotranspiration by irrigated crops or consumed by industrial processes, manufacturing, power generation, or municipal uses. Stream depletions include both beneficial and nonbeneficial consumptive uses.

The determination of historic consumptive use involves analysis of a number of factors, all of which are subjected to engineering judgment and legal interpretation. The first factor established is usually the study period, which is selected to represent historic conditions. This is the period of record to be analyzed and should be representative of the conditions under which the water rights were exercised. In selecting a study period, it is important that streamflow and climatological records be available for analysis and that the period contain at least one critically dry year. Records of recent years are probably more accurate and will reflect current administrative practices. Earlier records are often more representative of the extent of past irrigation, which has recently receded in many areas in the face of urbanization and other factors leading to the decline of irrigated agriculture.

The most common need for determining historic consumptive use involves an irrigation water right that is to be changed to some other time, type, or place of use. To do this without allowing an enlarged use or causing injury to other water rights means that both the quantity and timing of the consumptive use under historic exercise of the right must be determined. This involves defining the type of crops irrigated, the diversions available under the right when in priority, and the potential and actual irrigation and consumptive use occurring as a result of the irrigation.

The potential consumptive use is that which the crop would consume if a full supply of water were available to meet plant growth needs. The actual consumptive use is the amount of the available irrigation water consumed. Irrigation consumptive use is the amount of consumptive use supplied by irrigation water applied in addition to the natural precipitation which is effectively available to the plant. In some cases irrigation consumptive use may be supplied by natural subirrigation, which is usually not included in the amount of beneficial historic consumptive use available for transfer or conversion to other uses.

Figure 20 is a schematic representation of a stream and irrigation system which shows the various components that must be analyzed and quantified in determining historic consumptive use. The principal components displayed in Figure 20 are discussed by number.

1. *River flow* upstream of the point of diversion represents the physical supply available at the ditch headgate. Since surface runoff records are rarely available at the point of interest, it is necessary to establish the flow available by extrapolation from data for gages upstream or downstream or by correlation of records from hydrologically similar basins.
2. The amount *diverted for irrigation* is referred to as the stream headgate diversion and represents what is normally recorded by the state water official (water commissioner). This amount should not be confused with the farm headgate delivery which is the amount actually applied to the area irrigated and will be less than the stream headgate diversion by the net value of canal losses due to (4) *evaporation*, (8) *seepage* and (9) *bypasses*, and canal gains from (5) *precipitation* and (6) *inflow* from surface runoff.
3. *Undiverted river flow* bypasses the irrigation headgate and may be supplemented by (15) *groundwater contributions* which are positive in gaining streams and negative in stream reaches that lose flow to the groundwater system, (16) *surface runoff from nonirrigated lands*, (17) *industrial and municipal discharges*, and (18) *natural inflow* from tributaries.

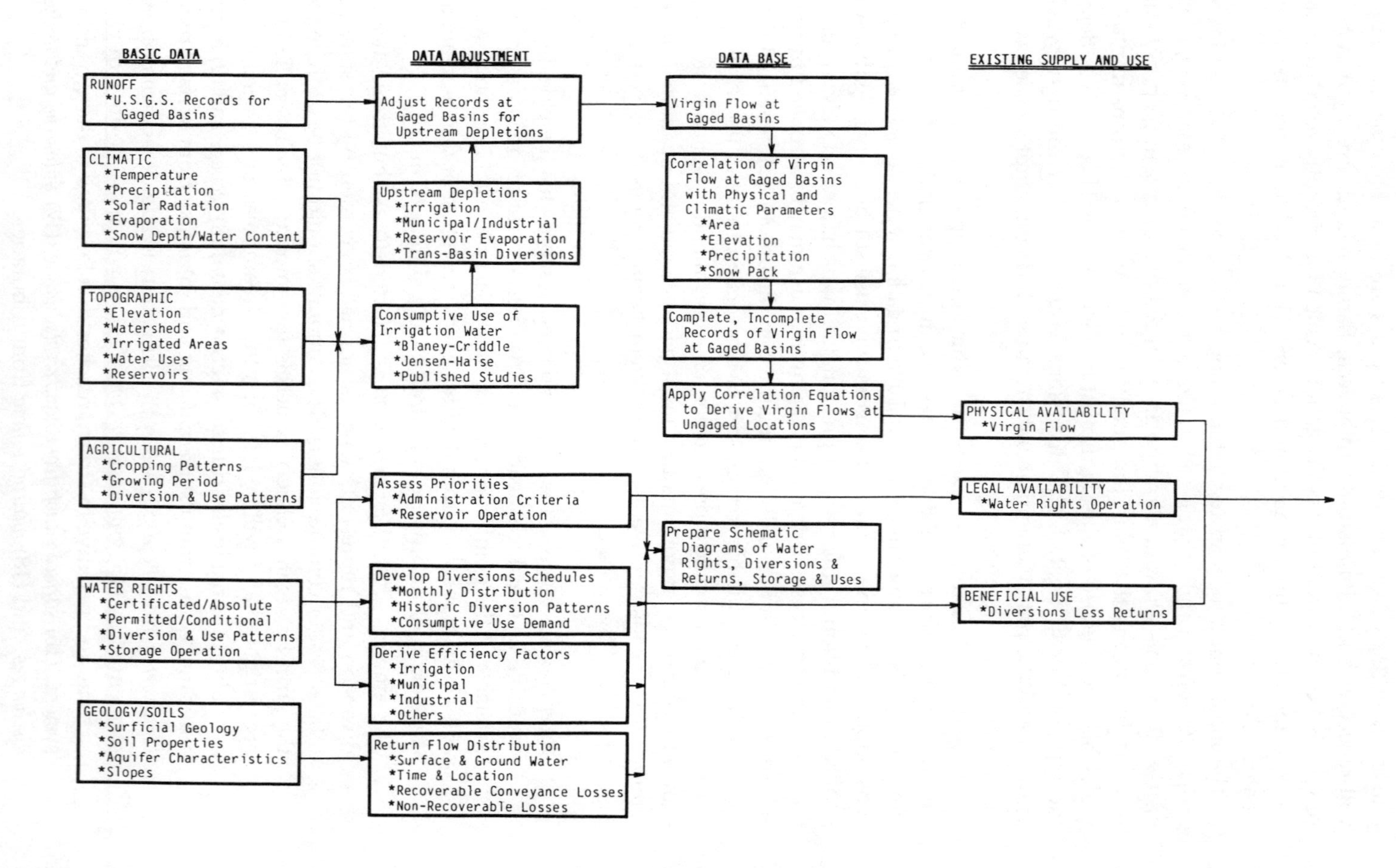
BASIC DATA
DATA ADJUSTMENT
DATA BASE
EXISTING SUPPLY AND USE
RUNOFF
*U.S.G.S. Records for Gaged Basins
CLIMATIC
*Temperature
*Precipitation
*Solar Radiation
*Evaporation
*Snow Depth/Water Content
TOPOGRAPHIC
*Elevation
*Watersheds
*Irrigated Areas
*Water Uses
*Reservoirs
AGRICULTURAL
*Cropping Patterns
*Growing Period
*Diversion & Use Patterns
WATER RIGHTS
*Certificated/Absolute
*Permitted/Conditional
*Diversion & Use Patterns
*Storage Operation
GEOLOGY/SOILS
*Surficial Geology
*Soil Properties
*Aquifer Characteristics
*Slopes
Adjust Records at Gaged Basins for Upstream Depletions
Upstream Depletions
*Irrigation
*Municipal/Industrial
*Reservoir Evaporation
*Trans-Basin Diversions
Consumptive Use of Irrigation Water
*Blaney-Criddle
*Jensen-Haise
*Published Studies
Assess Priorities
*Administration Criteria
*Reservoir Operation
Develop Diversions Schedules
*Monthly Distribution
*Historic Diversion Patterns
*Consumptive Use Demand
Derive Efficiency Factors
*Irrigation
*Municipal
*Industrial
*Others
Return Flow Distribution
*Surface & Ground Water
*Time & Location
*Recoverable Conveyance Losses
*Non-Recoverable Losses
Virgin Flow at Gaged Basins
Correlation of Virgin Flow at Gaged Basins with Physical and Climatic Parameters
*Area
*Elevation
*Precipitation
*Snow Pack
Complete, Incomplete Records of Virgin Flow at Gaged Basins
Apply Correlation Equations to Derive Virgin Flows at Ungaged Locations
Prepare Schematic Diagrams of Water Rights, Diversions & Returns, Storage & Uses
PHYSICAL AVAILABILITY
*Virgin Flow
LEGAL AVAILABILITY
*Water Rights Operation
BENEFICIAL USE
*Diversions Less Returns

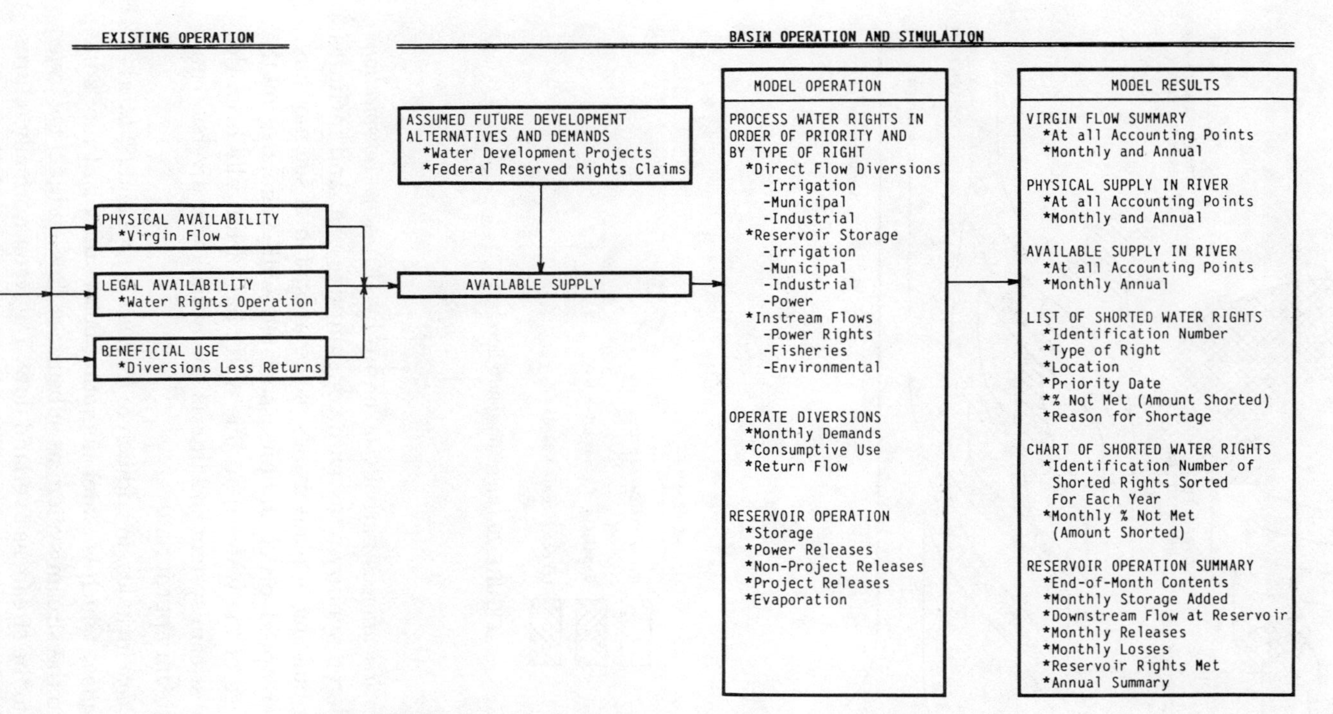


**FIGURE 19.** River basin water rights and supply operation model.

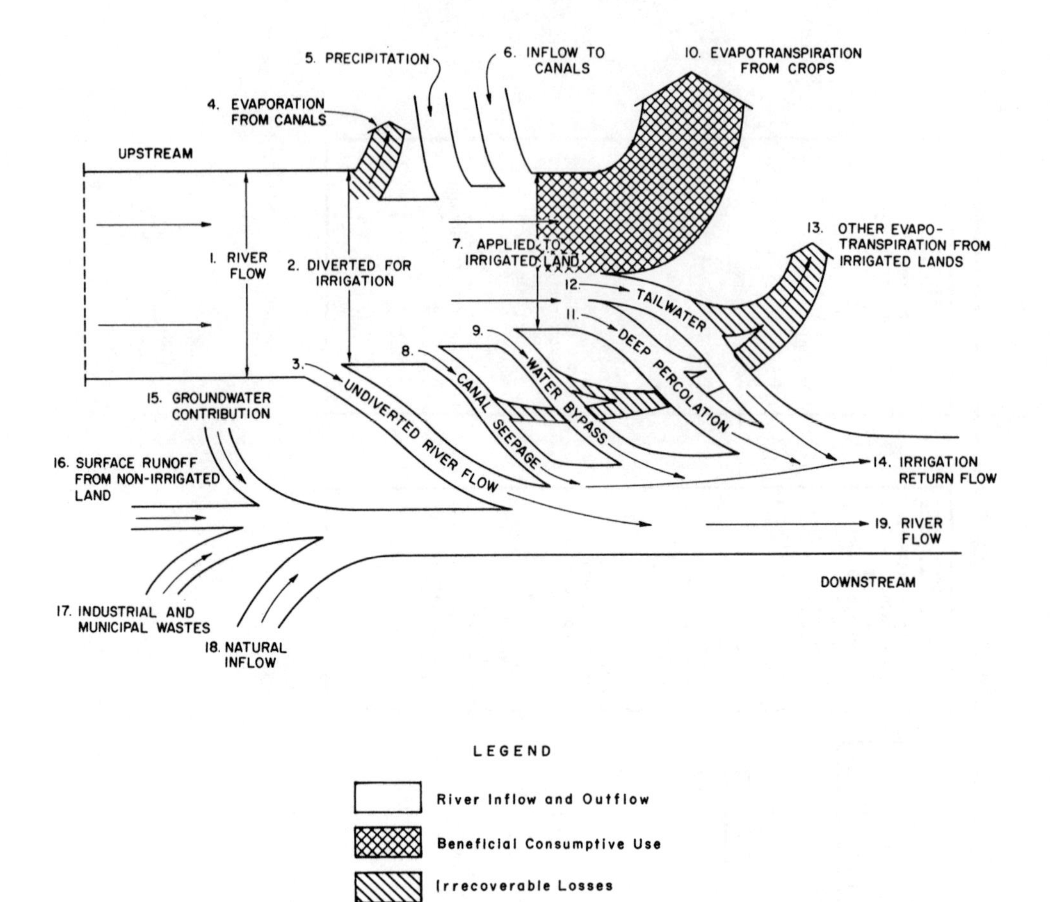


**FIGURE 20.** Irrigation return flow.

7. Water *applied to irrigation* is partially taken up by (10) *evapotranspiration,* usually considered synonymous with consumptive use, and includes transpiration or building of plant tissues plus evaporation of soil moisture, snow and intercepted precipitation associated with vegetal growth. Through exercise of the water right this water has historically been depleted from the stream system and thus is not available for diversion and use by downstream appropriators.

14. *Irrigation return flow* includes, in addition to (8) *canal seepage* and (9) *bypasses,* (11) *deep percolation,* which is water migrating below the plant root zone and returned to the stream system as subsurface flow and (12) *tail water* which returns to the stream as overland flow. Tail water normally returns to the stream within a matter of days, whereas deep percolation flows through the soil at a much slower rate, in many cases taking several months to reach the stream, thereby contributing significantly to winter flows.

13. *Other losses* to the atmosphere occur as a result of the irrigation operation, but in most cases are too small to be quantified individually and are grouped within the other major losses.

When evaluating the historic operation of irrigation water rights it is useful to calculate the irrigation efficiency by dividing the consumptive use by the amount diverted. The result for normal flood irrigation practice will range between 40 to 60 percent, meaning that 60 to 40 percent of the water diverted at the stream headgate returns to the stream. Other methods of irrigation using center pivot or linear sprinklers and drip irrigation systems have higher efficiencies, on the order of 80 to 95 percent. Table 4 shows how headgate diversions can be broken down into various kinds of conveyance and farm losses, including recoverable and irrecoverable losses, to arrive at the operating efficiency.

Diversions to a single ditch may be made under one or more separate decrees with different priorities and water from several ditches may be delivered to the same land. Ditches may also divert from more than one source and may carry both direct flow and storage water. This often creates problems in identifying and quantifying the land irrigated by individual water rights. Figure 21 illustrates a system of ditches and reservoirs used to irrigate lands on a ranch in western Colorado. For administrative purposes, it was necessary to determine and quantify the historic use of the water delivered by the McMahon Ditch from Red Dirt Creek to irrigated lands also served by other ditch and reservoirs, as shown. The McMahon Ditch had decrees from Red Dirt Creek and Deer Creek and also carried storage water released from McMahon Reservoir. The parcels could also be supplied by water diverted from Pinto Creek, through the Heini Ditch and Reservoir system and from Lewis Reservoir.

To resolve this problem, it was necessary to evaluate aerial photographs for six different years, covering a 45-year period, interview operators of the ditch and reservoir systems, the local water commissioner, and the managers responsible for irrigating the various parcels. Computations were made of the consumptive use of the irrigated lands, which, when applied to irrigation efficiencies, indicated the quantity of water needed at the McMahon Ditch, Red Dirt Creek, and Deer Creek headgates. This information was compared to the natural flow available, after satisfaction of obligations to downstream senior water rights, based on estimates from an existing surface runoff gage on Red Dirt Creek and estimates of the flows in Deer Creek and Pinto Creek derived by extrapolation from similar basins, to derive the amount of water that had historically been supplied from McMahon Reservoir. The results of an investigation such as this must be thoroughly documented and substantiated for use in negotiation with other water users and state officials or presentation to an administrative or judicial proceeding.

The unit consumptive use of irrigation water (volume of consumptive use per unit of area, commonly expressed as acre-feet per acre or simply feet or inches) by crops can be measured or computed. Measurement is accomplished by instruments called lysimeters, which are tanks filled with soil in which crops can

Table 4
DISTRIBUTION OF LOSSES AND RETURN FLOWS

| Function | Loss Expressed as a Percentage of: | | Distribution of Nonrecoverable Losses and Recoverable Return Flow for an Assumed Headgate Diversion of 100 af | | | | | |
|---|---|---|---|---|---|---|---|---|
| | | | Nonrecoverable | | Recoverable | | Total | |
| | Headgate Diversion | Indicated Function | Percent | Amount (af) | Percent | Amount (af) | Percent | Amount (af) |
| Conveyance | 25 | | | | | | | |
| Surface | | 25 | 30 | 1.88 | 70 | 4.38 | 100 | 6.25 |
| Seepage | | 75 | 5 | 0.94 | 95 | 17.81 | 100 | 18.75 |
| Subtotal Conveyance | | | | 2.81 | | 22.19 | | 25.00 |
| Crop Evapotranspiration | 38 | 100 | 100 | 38.00 | 0 | 0 | 100 | 38.00 |
| On Farm Application | 37 | | | | | | | |
| Surface | | 50 | 15 | 2.78 | 85 | 15.73 | 100 | 18.50 |
| Percolation | | 50 | 5 | 0.93 | 95 | 17.58 | 100 | 18.50 |
| Subtotal On Farm | | | | 3.70 | | 33.30 | | 37.00 |
| Total Losses | | | | 44.51 | | 55.49 | 100 | 100.00 |
| Surface | | | | 4.65 | | 20.10 | | 24.75 |
| Subsurface | | | | 1.86 | | 35.39 | | 37.25 |
| Ditch Headgate Efficiency | 38% | | | | | | | |
| Farm Headgate Efficiency | 51% | | | | | | | |

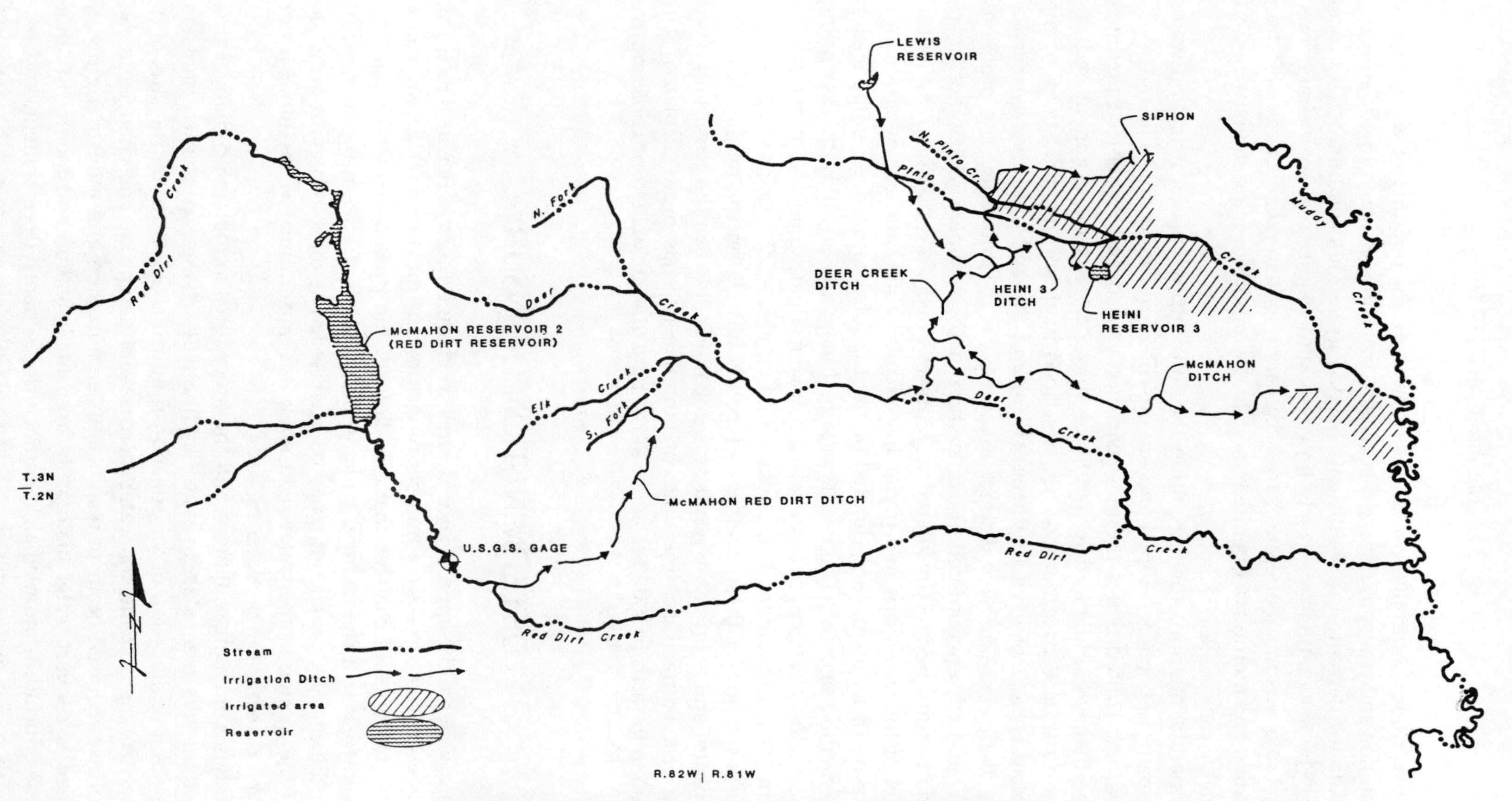


**FIGURE 21.** Western Colorado Irrigation ditch and reservoir system.

be grown under controlled conditions to measure the quantity of water lost by evaporation and transpiration. Measurement by lysimeters can provide site-specific data for deriving the coefficients needed in various computational procedures. The installations are costly however, and require regular maintenance. At least two seasons of operation are necessary to obtain reliable data. Selection of location and method of operation are equally important to provide usable and acceptable data.

Numerous methods for computing the consumptive use of irrigation water by crops have been developed and are described in the technical literature. In the western United States the most commonly used and recognized methods are the Blaney-Criddle and Jensen-Haise formulas. The development and application of these two methods are described in detail in publications of the American Society of Civil Engineers (1973) and the Soil Conservation Service (1970). Both methods have distinct advantages and limitations. The problems most commonly encountered in using either method involve selection of an appropriate study period considered to represent historic conditions, the identification of the crops irrigated under historic operation, and the determination of crop coefficients, all compounded by the lack of data needed for application of the method selected. The results obtained may vary significantly depending on the method of computation selected, even when identical parameters are used in the computation. This is illustrated by Figure 22.

The resolution of these problems depends heavily on the judgment of the investigator and is often the result of negotiations between the parties involved in the proceeding. Once the unit consumptive use has been calculated, it is applied to the area irrigated to arrive at the total volume of historic consumptive use.

## DETERMINATION OF INJURY

When a water right appropriator desires to change the manner in which the right is exercised there are a number of factors that must be considered. These include the types of changes that are allowed, the procedures which must be followed to obtain the change, and the principles that govern the manner in which the changed water right may be exercised. The most important principle is that the change of the water right must not cause injury to any other water right, particularly junior water rights.

The types of changes allowed and the procedures for obtaining changes in a water right vary from state to state. Traditionally, changes concerned transfers of the point of diversion of an irrigation right from one place on the stream to another. As the demand for water increases in magnitude and changes in the type of use required occur in response to urbanization and industrial growth, the need to convert water historically used for irrigation and mining to other purposes, including municipal, manufacturing, energy development, and recreation, becomes more common. The state of Colorado recognized this need and

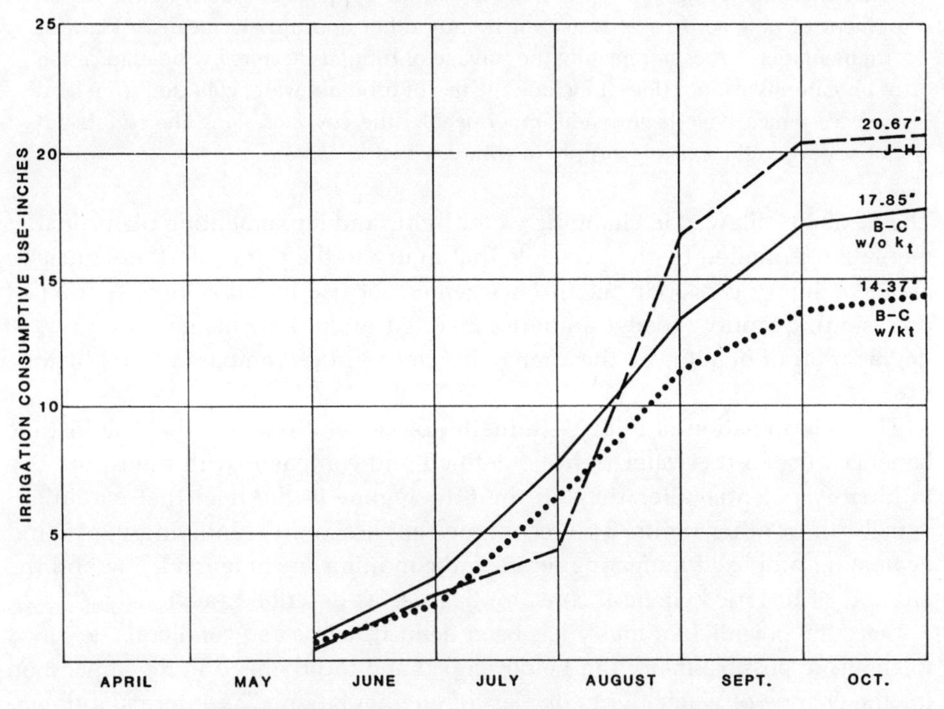


**FIGURE 22.** Variation in computations of consumptive use.

in 1969 enacted the Water Rights Determination and Administration Act, which defined a change of water right as follows:

> "Change of water right" means a change in the type, place, or time of use, a change in the point of diversion, a change from a fixed point of diversion to alternate or supplemental points of diversion, a change from alternate or supplemental points of diversion to a fixed point of diversion, a change in the means of diversion, a change in the place of storage, a change from direct application to storage and subsequent application, a change from storage and subsequent application to direct application, a change from a fixed place of storage to alternate places of storage, a change from alternate places of storage to a fixed place of storage or any combination of such changes. The term "change of water right" includes changes of conditional water rights as well as changes of water rights.

This broad definition recognized that the expeditious use of water necessitates flexibility in changing water rights corresponding with the need to maximize beneficial use of the state's water resources. This was emphasized by the provisions for establishing a plan of augmentation which is defined below:

> "Plan for Augmentation" means a detailed program to increase the supply of water available for beneficial use in a division or portion thereof by the development of new or alternate means or points of diversion, by a pooling of water resources, by water exchange projects, by providing substitute supplies of water, by the development of new sources of water, or by any other appropriate means. "Plan for Augmentation" does not include the salvage of tributary waters by the eradication of phreatophytes, nor does it include the use of tributary water collected from land surfaces which have been made impermeable, thereby increasing the runoff but not adding to the existing supply of tributary water.

The flexibility allowed in changing water rights and implementing plans of augmentation is limited by the principle that injury to the rights of others must be avoided. Injury can occur as an enlargement of use in either time (period of diversion), quantity (greater amounts diverted under the changed priority), or degradation of quality, by the changed water right as compared to its historic use.

The determination of injury is a matter of fact and requires that the historic beneficial use of the water right be defined and compared with the future use to identify potential alterations in the flow regime of the river that would adversely affect other rights. The techniques applicable to identifying injury and evaluating proposed mitigating terms and conditions include modeling and the analysis of historic beneficial consumptive use as described previously.

Once the potential for injury has been defined, terms and conditions designed to eliminate possible injury can be developed and incorporated in the application for the change of water rights or plan of augmentation. These terms and conditions become negotiable between the applicant seeking the change and the objectors to the change and are included in the decree granting the change of water rights or the plan of augmentation. The general nature of terms and conditions that may be proposed is stated in the 1969 act as follows:

> Terms and conditions to prevent injury may include:
>
> 1. A limitation on the use of the water which is subject to the change, taking into consideration the historic use and the flexibility required by annual climatic differences;
> 2. The relinquishment of part of the decree for which the change is sought, or the relinquishment of other decrees owned by the applicant which are used by the applicant in conjunction with the decree for which the change has been requested, if necessary to prevent an enlargement upon the historic use or diminution of return flow to the detriment of other appropriators;
> 3. A time limitation on the diversion of water for which the change is sought in terms of months per year;

4. Such other conditions as may be necessary to protect the vested rights of others.

The nature and extent of potential injury liable to result from a proposed change of water rights varies with the type of change sought. Table 5 summarizes the principal types of water right changes encountered, with examples, and the corresponding nature of potential injury and some commonly applied terms and conditions for mitigation. A change of water rights or plan of augmentation will usually involve one or more types of change and may require the imposition of a combination of the terms and conditions listed in Table 5, plus other conditions limited only by the imagination of the applicants and objectors. In some cases, provisions may be included in the decree for monitoring of the changed water rights or plan of augmentation over a period of years to ensure the terms and conditions do prevent the injury anticipated.

The factual determination of potential injury and the definition of appropriate terms and conditions should be based on analysis of historical consumptive use under average and dry year conditions. This includes determination of the quantity, timing, and location of return flows, the extent to which the rights involved have historically been exercised in priority, and the degree to which junior rights have been dependent upon the availability of return flow from the rights to be changed or augmented. It is also necessary to consider the possibility that the change will increase the frequency of junior rights being called out by the changed right.

This last consideration requires analysis of the relative location and priority of water rights on the stream system involved, usually facilitated by means of a straight-line diagram as illustrated by Figure 23, and an awareness of the possibility for rebound calls, which occur when a transferred senior is, as a result of the transfer, in a position to call out a junior not previously affected. This call may then rebound from the newly affected juniors who, as a result, will be called out more frequently than before the change and thereby suffer a diminishment of their supply.

The terms and conditions proposed in a change of water rights or plan of augmentation must be administrable by the state engineer. When negotiating and specifying terms and conditions, the parties involved in the proceeding must recognize the limits of water commissioners and others in making measurements and recording data. There are cases where the final document became so complex, in terms of monitoring, data collection, and administrative requirements, that the applicant was required to pay the state the cost of an additional water commissioner required to administer the plan.

## Example of Change in Place of Storage and Type of Use

In 1955, the city of Broomfield acquired title to storage decrees in the Zang Reservoir No. 1, Zang Reservoir No. 2, and Nissen Reservoir No. 6 from a development corporation. The development corporation acquired title to land that is

Table 5
WATER RIGHTS CHANGES
SOURCES OF INJURY AND MITIGATING MEASURES

| Type of Change | Example | Source of Injury | Terms and Conditions |
|---|---|---|---|
| Type of Use | Irrigation to municipal/industrial. | Extension of diversion period from seasonal to year-round. | Limit diversions to historic irrigation season. Limit seasonal volume diverted to historic consumptive use. |
| | | Elimination, reduction, or alteration of return flow historically available to downstream junior appropriator. | Return a portion of the water available for use by the changed right to the stream when diverted to maintain historic conditions. |
| | Mining to irrigation recreation or municipal. | Increase in consumptive use (depletion). | Abandon portion of right to stream. |
| | | Degradation of quality. | Treatment of effluent. |
| Place of Use | Transfer point of diversion of ditch along river. Transfer ditch priority to well(s) as alternate points of diversion. | Increase in period and quantity of diversion due to greater availability of water at new or alternate point of diversion. | Limit diversions at new point or wells to periods when water physically available and in priority at original point of diversion. Assess stream conveyance losses against diversions at new point of diversion or wells. |

| | | | |
|---|---|---|---|
| Time of Use | Irrigation to snowmaking. | Diversion for snowmaking is 100% depletion in fall with return in spring reduced by losses to evaporation and sublimation and delayed by groundwater return portion. | Provision of replacement water from other sources, such as nontributary wells or imported water.<br>Subordination of changed right priority to downstream junior to insure juniors supply not diminished by call from senior placed outside of historic irrigation season. |
| Direct flow to storage, usually accompanied by a change in type, place, or time of use. | Direct flow irrigation right changed to storage for municipal or industrial use.<br>Mining right stored for use as source of augmentation water. | Alteration of historic return flow available to downstream junior.<br>Enlarged use and increased depletion due to difference in consumptive uses and timing of diversions and returns. | Limit amount stored to historic consumptive use.<br>Require releases from storage to compensate for lost return flows.<br>Limit period when water can be diverted to storage. |

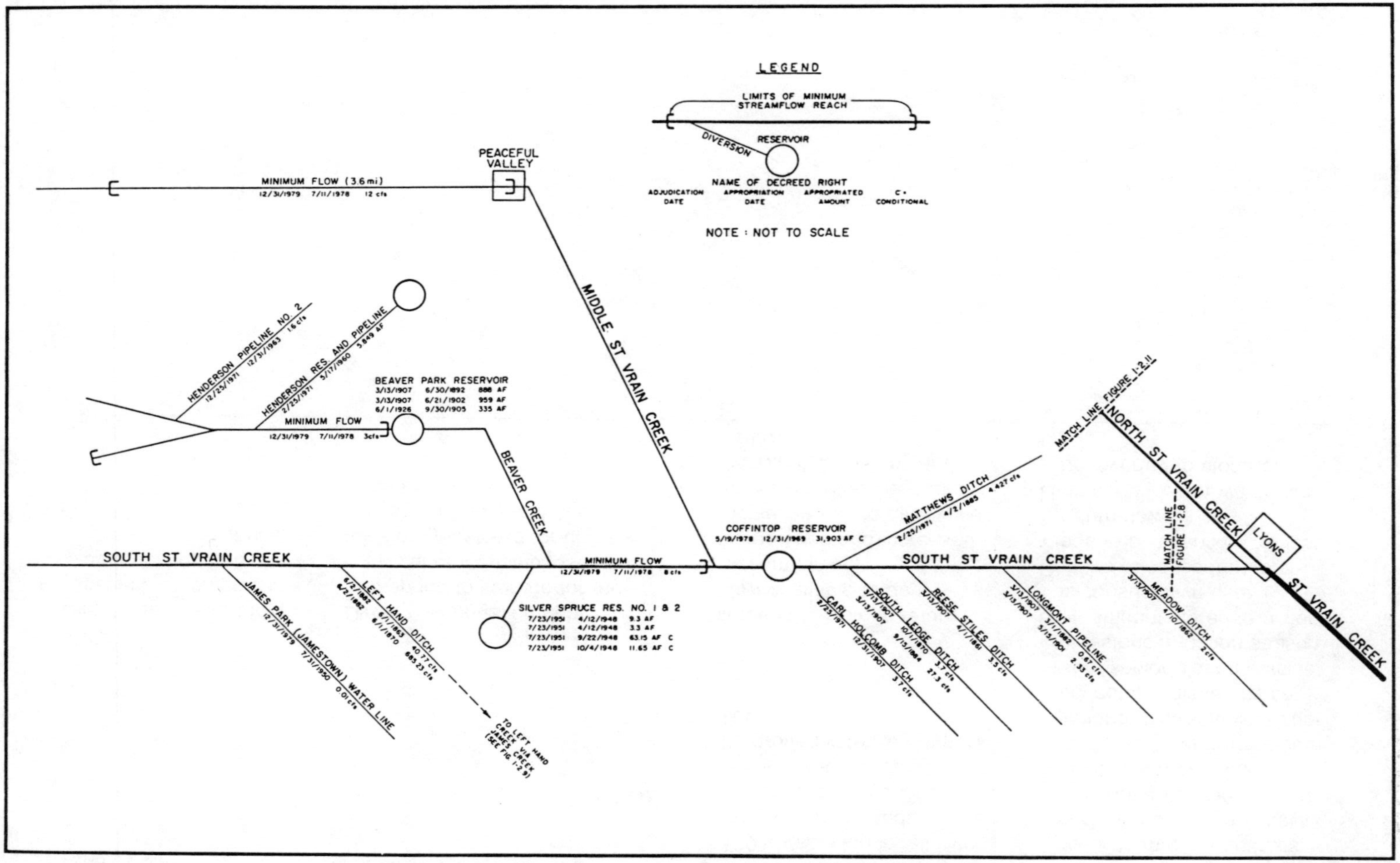

LEGEND

LIMITS OF MINIMUM STREAMFLOW REACH

DIVERSION

RESERVOIR

NAME OF DECREED RIGHT

ADJUDICATION DATE | APPROPRIATION DATE | APPROPRIATED AMOUNT | C = CONDITIONAL

NOTE : NOT TO SCALE

PEACEFUL VALLEY

MINIMUM FLOW (3.6 mi)
12/31/1979 7/11/1978 12 cfs

MIDDLE ST VRAIN CREEK

HENDERSON PIPELINE NO. 2
12/23/1971 12/31/1963 1.6 cfs

HENDERSON RES AND PIPELINE
2/23/1971 3/17/1960 5849 AF

BEAVER PARK RESERVOIR
3/13/1907 6/30/1892 888 AF
3/13/1907 6/21/1902 959 AF
6/1/1926 9/30/1905 335 AF

MINIMUM FLOW
12/31/1979 7/11/1978 3cfs

BEAVER CREEK

SOUTH ST VRAIN CREEK

JAMES PARK (JAMESTOWN) WATER LINE
12/31/1979 7/31/1950 0.01 cfs

LEFT HAND DITCH
6/2/1882 6/1/1863 40.77 cfs
6/2/1882 6/1/1870 685.23 cfs

TO LEFT HAND CREEK VIA JAMES CREEK (SEE FIG I-2.9)

MINIMUM FLOW
12/31/1979 7/11/1978 8 cfs

SILVER SPRUCE RES. NO. 1 & 2
7/23/1951 4/12/1948 9.3 AF
7/23/1951 4/12/1948 3.3 AF
7/23/1951 9/22/1948 63.15 AF C
7/23/1951 10/4/1948 11.65 AF C

COFFINTOP RESERVOIR
5/19/1978 12/31/1969 31,903 AF C

MATTHEWS DITCH
2/25/1971 4/2/1883 4.427 cfs

CARL HOLCOMB DITCH
2/25/1971 12/31/1907 3.7 cfs

SOUTH LEDGE DITCH
3/13/1907 10/1/1870 3.7 cfs
3/13/1907 9/15/1884 27.3 cfs

REESE STILES DITCH
3/13/1907 4/1/1861 3.5 cfs

LONGMONT PIPELINE
3/13/1907 3/1/1882 0.67 cfs
3/13/1907 5/15/1901 2.33 cfs

MEADOW DITCH
3/13/1907 4/10/1882 2 cfs

SOUTH ST VRAIN CREEK

MATCH LINE FIGURE I-2.11

NORTH ST VRAIN CREEK

MATCH LINE FIGURE I-2.8

LYONS

ST VRAIN CREEK

**FIGURE 23.** Straight-Line diagram.

now downtown Broomfield and proceeded to develop the land occupied by the reservoir and irrigated by water from the three reservoirs and an irrigation ditch. Between 1955 and 1983, the city of Broomfield obtained its raw water supply by diverting directly out of an irrigation ditch and pumping the water into it's Great Western Reservoir. Water from the reservoir was delivered to the city's water treatment plant and distribution system by gravity. In 1982, the city applied for a change in the place of storage from the three reservoirs to the Great Western Reservoir and a change in use from irrigation to municipal purposes. Several objectors filed protests citing three principal issues:

1. Abandonment, since no record of use had been made since 1955.
2. Injury to junior water rights due to an expansion of use if the three senior reservoir priorities were exerted against appropriations made subsequent to 1955.
3. Loss of return flows historically accruing to the stream system from the use of water for irrigation purposes.

Aerial photographs taken in 1937 and 1941 were used to quantify the area irrigated by the original appropriators and subsequent users of the Zang and Nissen storage decrees.

Interviews with resident farmers provided representative cropping patterns that were used to compute the consumptive use under average, dry and wet year conditions. The consumptive use was converted to a farm headgate diversion by applying a farm headgate efficiency based on local practice and interviews with the former farmers of the land. The farm headgate diversion required to support the documented historic irrigation was then compared to the quantity of water available to the property from all sources, including direct flow ditch rights and storage in Zang and Nissen reservoirs.

This analysis demonstrated that the water stored in Zang and Nissen reservoirs was essential to support the historic level of irrigation and established the historic use of the reservoir decrees. These data were used to establish that the rights had not been abandoned and to define terms and conditions that would allow the city to utilize its storage decrees, while at the same time protecting vested water rights. The following terms and conditions were adopted:

1. Sixty-five percent of the total amount decreed to the reservoirs was transferred to Great Western Reservoir and changed to municipal use.
2. Thirty-five percent of the total amount decreed was abandoned to the stream to account for historic return flows.
3. Diversions under the Zang and Nissen priorities were limited to the period April 21st to August 1st to protect junior rights.
4. The maximum rate of diversion under the decrees was limited to 40 cfs, which represented 58 percent of the originally decreed rate of diversion.

## DILIGENCE

An important principle of the appropriation doctrine is that the validity of an appropriation depends on its being perfected within a reasonable time by the exercise of due diligence. The basis for this principle allows for the development of large projects requiring long periods of time for completion by providing for the priority of the right to relate back to the date of the first step taken to initiate the appropriation. To prevent speculation and encourage development, the appropriator must demonstrate due diligence in perfecting the right and applying the water to beneficial use. The procedures and requirements for demonstrating due diligence to state water authorities or water courts varies from state to state.

The definition of what constitutes due diligence is a question of fact that will vary with the circumstances of each project. The facts to be considered include the specific steps taken in perfecting the right, the elapsed time between actions, delays in project development caused by avoidable or unavoidable circumstances, and the magnitude of the project. The Colorado Supreme Court describes the factors which should be considered in determining due diligence as follows:

> The question of diligence must be determined in light of all factors present in a particular case, including the size and complexity of the project; the extent of the construction season; the availability of materials, labor, and equipment; the economic ability of the claimant; the intervention of outside delaying factors, such as wars, strikes, and litigation.
> *(Colorado River Water Conservation District v. Twin Lakes Reservoir and Canal Co.*, 1970)

From a factual standpoint, diligence in developing a project and perfecting a water right can be demonstrated by application of the normal planning, design, and construction process. This includes data collection, planning studies, and project formulation and design. Table 6 outlines these activities and provides a guide to actions that can be undertaken in a progressive manner to diligently develop a water resource project. The activities selected for due diligence should be as site-specific to the project as possible. Complete records of the cost incurred should be maintained and allocated to specific projects.

## WATER RIGHTS APPRAISAL

The value of a water right is a function of the amount of water that can be derived by exercise of the right, the dependability of the supply, the uses to which water can be applied, and when and where the water will be available to the purchaser. These factors are directly related to the type of decree and the physical and legal availability of water.

Table 6
OUTLINE OF DILIGENCE ACTIONS

- I. Data Collection
  - A. General
    - 1. Water resources
      - a. Hydrology: precipitation, streamflow, water quality, evapotranspiration, sediment
      - b. Geology: groundwater, soil classification, erosion, structure
      - c. Cartographic: topographic, geologic, soils, land ownership, and cultural mapping
    - 2. Other basic resources
      - a. Geologic: minerals, embankment, and borrow materials
      - b. Ecologic: vegetation, fish, and wildlife
      - c. Demographic: people and institutions
      - d. Economic: industry, transportation, markets, tourism, recreation, land, taxes
    - 3. Constraints
      - a. Legal: water rights, pollution control, zoning, land ownership, administrative requirements, Indian lands, interstate compacts, treaties
      - b. Public opinion
      - c. Existing projects
  - B. Special data
    - 1. Agriculture
      - a. Land classification
      - b. Crop water requirements: quantity and quality
      - c. Climatic limitations
    - 2. Municipal uses
      - a. Industrial water needs: quantity and quality
      - b. Population water needs
    - 3. Hydropower
      - a. Projected needs
      - b. Alternate sources
    - 4. Flood mitigation
      - a. Extent of past flooding and damages
      - b. Local storm drainage requirements
    - 5. Recreation
      - a. Natural attractions
      - b. Present recreation patterns: types, place, time
    - 6. Pollution control
      - a. Existing waste discharges: location, time, character of waste
      - b. Water pollution regulations or quality standards

(*continued*)

Table 6 (*continued*)

II. Projections
  A. General
    1. Population: place and time
    2. Land use: place and time
    3. Economic: markets, tourism
  B. Specific
    1. Agriculture
      a. Markets
      b. Crops
      c. Technological development
      d. Water demand
    2. Municipal
      a. Domestic water demand
      b. Industrial water demand
      c. Technological changes
    3. Power
      a. Market and demand
      b. Growth of alternative sources
      c. Technological improvements
    4. Flood Control
      a. Zoning possibilities
      b. Projected flood damage: place and frequency
      c. Possibilities of flood warning
    5. Recreation
      a. Growth of demand
      b. Changes in recreation preferences
    6. Pollution
      a. Anticipated quantities and characteristics of waste
      b. Technological advances

III. Project Formulation
  A. Defining boundary conditions
  B. Listing all possible land-use plans and their water requirements
  C. Listing all possible projects which could meet the needs projected
  D. Preliminary designs and cost estimates
  E. Estimating benefits
  F. Reports on final alternatives showing costs, benefits, staging, financing, and intangible factors for review by project developer

IV. Construction
  A. New or extended ditches, canals, or other conveyance structures.
  B. Diversion works
  C. Gages, flumes, or meters
  D. Access roads, construction housing or offices
  E. Leveling and grading of fields
  F. Wells and sprinkler systems

In most western states, water rights are treated as real property and may be purchased and sold separately from the land and the type, time, and place of use changed provided that other vested water rights are not adversely affected. An administrative or judicial proceeding is usually required to convert water rights from one use to another. Certain limits may be imposed upon the new use to protect other vested rights. Such limits often reduce the amount of water available to the new use. Because of the uncertainties involved in the change proceeding, it is difficult to predict how much water will be available to a purchaser and in many cases, water rights are purchased on the basis of a unit price per acre-foot of water actually obtained by the buyer at his point of use over a specified period of time after the change is granted.

Historic consumptive use is the most generally accepted measure of the amount of water available for sale and conversion under a water right. It represents that quantity of water historically consumed by beneficial use of the right and therefore not available to other water users on the stream. The determination of the historic consumptive use involves an analysis of the crops irrigated, amount of water diverted, return flow patterns, and other factors.

Once the historic consumptive use has been determined, the value of the water rights can be established in terms of a unit value per acre-foot of consumptive use. The legal and administrative characteristics of a water right have a significant influence of its value. Native water, that is, water originating in the basin of use, is administered by priority of appropriation and is generally limited to one use only. Imported and developed surface water and nontributary groundwater are considered free of administration by priority if they can be identified and accounted for. This means that such water can be used to augment out of priority depletions, used by exchange, successively used for different purposes or sold, leased, or otherwise disposed of after its initial use. Thus, a much higher value attaches to a water right for imported or developed surface water or nontributary groundwater.

Values are usually derived by examining recent sales of a similar nature involving arm's-length transactions between knowledgeable buyers and knowledgeable sellers. Because there is no organized continuous market for water rights, as there is for land, it is difficult to document transactions. Those that are documented reflect wide variations in unit prices. These variations occur because different terms are often used to measure water rights and different values are placed on priority, location, and past operation. The value of the purchased water is influenced by its proposed use.

Various units of measurement are used to describe water rights. These include the allowable rate of diversion in cubic feet per second (cfs), the total quantity diverted during a season in acre-feet and the consumptive use in acre-feet. The yield of a water right may be defined in terms of diversions from the stream or delivery to the crop after conveyance losses. Yield and consumptive use may also be expressed in terms of average or minimum values related to a specific period of time.

Storage rights are generally measured in terms of the amount of water that can be stored on a dependable basis. Storage rights are usable throughout the year and provide a means of regulating erratic natural runoff and variable direct flow diversions from month to month and year to year. For this reason, storage rights are often assigned a higher value than direct flow rights, unless the direct flow rights are senior enough to produce a highly dependable yield.

# GLOSSARY

**Abandonment:** The loss of a water right based on the coexistence of the intent not to use the right with actual nonuse of that water right.

**Absolute Water Right:** A water right that has been perfected and placed to beneficial use.

**Acre-Foot:** The volume of water required to cover 1 acre of land to a depth of 1 foot; 325,850 gallons or 1233.5 cubic meters. One acre-foot supplies a family of four for about one year.

**Adjudication:** A judicial proceeding in which a priority is assigned to an appropriation and a decree issued defining the water right.

**Administrative Procedures:** Proceedings before an officer of the executive branch of government as distinguished from proceedings before the judicial branch of government.

**Alluvium:** Deposits of sand and gravel derived from erosional processes and laid down in river channels and floodplains.

**Appropriation:** The diversion of a certain portion of the waters of the state and the application of same to a beneficial use (under certain conditions an appropriation for instream flow or minimum lake level maintenance may be accomplished without the act of diversion and application to beneficial use).

**Aquifer:** A water bearing formation.

**Aquifer Constant (alpha):** A number characteristic of an aquifer which denotes the speed with which transient changes will take place within the aquifer. It is represented by the symbol $\alpha$.

**Artesian Well:** A well that taps a confined aquifer and may have a pressure sufficient to support a flowing well.

**Artificial Recharge:** The addition of water to the groundwater reservoir by human activities, such as irrigation or induced infiltration from streams, wells, or spreading basins.

**Bank Storage:** The water contained in an aquifer hydraulically connected with a stream or lake and capable of supplying water to the stream or lake following a lowering of the free water surface or storing water flowing from the stream or lake on a rise of the free water surface.

**Barrier:** An impermeable formation in contact with an aquifer which confines the flow of groundwater to the aquifer.

**Basin Rank:** A number used in Colorado by the state engineer in the tabulation of decreed water rights to indicate the relative standing of a decreed right with respect to all other decreed rights within a water division.

**Beneficial Use:** The use of that amount of water that is reasonable and appropriate under reasonable efficient practices to accomplish, without waste, the purpose for which the diversion is lawfully made and without limiting the generality of the foregoing, shall include impoundment of water for recreational purposes, including fishery or wildlife.

**Boundary Conditions:** Conditions imposed by boundaries.

**Call:** The placing of a call by a senior priority to the water commissioner to shut down junior priorities so that the senior is able to divert its full entitlement. In such cases, junior priorities are curtailed or called out.

**Cienaga:** 1. An area where the water table is at or near the surface of the ground. Standing water occurs in depressions in the area, and it is covered with grass or sometimes with heavy vegetation. The term is usually applied to areas ranging in size from several hundred square feet to several hundred or more acres. Sometimes springs or small streams originate in the cienaga and flow from it for short distances. 2. An elevated or hillside marsh containing springs. Local in Southwest.

**Compact:** A contract between states of the union, entered into with the consent of the national government, and in water, defining the relative rights of two or more states on an interstate stream to use the waters of that stream.

**Condition of Continuity:** The requirement that water volumes must be strictly accounted for.

**Conditional Water Right:** A right to perfect a water right with a certain priority upon the completion with reasonable diligence of the appropriation upon which such water right is to be based.

**Cone of Depression:** The resulting water table form representing the gradient towards a well caused by withdrawals from the aquifer.

**Confined Aquifer:** An aquifer enclosed between impermeable formations.

**Consistent Units:** A unit that permits only one unit of a kind. Computational procedures require units be consistent. Data expressed in units other than those of a chosen system must be converted to the chosen system. Some conversion factors used in groundwater computations are given below:

| To Convert | To | Multiply by |
|---|---|---|
| Gallons per minute | Cubic feet per second | 0.002228 |
| Meinzers unit (permeability) | Feet per second | $1.5472 \times 10^{-6}$ |
| Meinzers unit (transmissivity) | Feet squared per second | $1.5472 \times 10^{-6}$ |
| Acre-feet | Cubic feet | 43,560 |
| Cubic feet per second | Gallons per minute | 448.8 |
| One year (365 days) | Seconds | 31,536,000 |
| One month (1/12 year) | Seconds | 2,628,000 |
| One day | Seconds | 86,400 |

A township has an area of 23,040 acres or $1003.62 \times 10^6$ square feet. A section has an area of 640 acres or $27.8784 \times 10^6$ square feet. One cubic foot per second running for one day will deliver 1.983471 acre-feet. One cubic foot per second running for 365 days will deliver 723.9669 acre-feet.

Note that while a year of 365 days is assumed for computation purposes, a year is 365.2422 days (Smithsonian Physical Tables). This is 31,556,930 seconds. A cubic foot per second running for one year will deliver 724.447 acre-feet.

**Consumptive Use:** The amount of water consumed during use of the water and no longer available to the stream system. For irrigation, consumptive use is water used by crops in transpiration and building of plant tissue.

**Conveyance Loss:** The loss of water from a conduit due to leakage, seepage, evaporation, or evapotranspiration.

**Creek:** A natural stream of water, normally smaller than, and often tributary to, a river.

**Critical Year:** Usually considered a year in which the annual precipitation was considerably less than average and runoff in most of the streams was low. The critical year is used to test the dependability of water rights under worst-case conditions.

**Darcy's Law:** A law discovered by Henry Philibert Gaspard Darcy (1803–1858). His experiments showed that the velocity of flow through porous media is proportional to the first power of the gradient.

**Decree:** An official document issued by the court defining the priority, amount, use, and location of a water right or plan of augmentation. When issued, the decree serves as a mandate to the state engineer to administer the water rights involved in accordance with the decree.

**Deep Percolation:** The drainage of soil water by gravity below the maximum effective depth of the root zone.

**Dependable Yield:** See Yield, firm.

**Depletion:** Net rate or quantity of water taken from a stream or groundwater aquifer and consumed by beneficial and nonbeneficial uses. For irrigation or municipal uses, the depletion is the headgate or well-head diversion less return flow to the same stream or groundwater aquifer.

**Developed Water:** Water so situated that it would not, but for human actions, contribute materially to either a natural stream or to nontributary groundwater, but is placed under control by some such artificial works as a mine or a tunnel.

**Diffuse Surface Water:** Water on the surface of the earth outside of a defined channel, typically flooding or urban runoff.

**Direct Flow Right:** A right defined in terms of discharge and which must be put to use more or less promptly following diversion from the source.

**Discharge, or Rate of Flow:** The volume of water passing a particular point in a unit of time. Units of discharge commonly used include cubic feet per second (cfs) or gallons per minute (gpm).

**Ditch:** A narrow trench cut into the surface of the ground to transport water from a stream to a point of use away from the stream.

**Divert:** Removing water from its natural course or location, or controlling water in its natural course or location, by means of a ditch, canal, flume, reservoir, bypass, pipeline, conduit, well, pump, or other structure or device.

**Diversion Records:** Record of the daily flow in cubic feet per second for a ditch or other diversion structure. Compiled by the district water commissioner, ditch rider, or other water official, diversion records are generally on file and available for review at the state engineer's office.

**Division Engineer:** The subordinate officers under the state engineer for each water division of the state, who perform the functions of the state engineer in those water divisions.

**Drainable Depth:** A depth of groundwater above the level of a system of drains or above the level of a river, in an aquifer hydraulically connected thereto.

**Drawdown:** The amount a water table has sunk from an initial stable configuration.

**Drift:** A term used to describe a flow of groundwater under the action of a naturally existing regional gradient.

**Due Diligence:** The effort necessary to bring an intent t to appropriate into fruition by the actual application of water to the beneficial use intended. Due diligence does not require unusual effort or expenditures, but only such constancy in the pursuit of the undertaking as is usual with those in like enterprises. Actions which demonstrate a good faith intention to complete the undertaking within a reasonable time.

**Dupuit-Forchheimer Idealization:** An idealization whose use was pioneered by Arsene Jules Emile Juvenal Dupuit (1804–1866) and Philipp Forchheimer (1852–1933). Under this idealization the gradient of the water table is assumed to be effective throughout the saturated thickness of the aquifer. When the water table gradient is small compared to unity, the postulated conditions are substantially realized.

**Duty of Water:** The total volume of irrigation water required to mature a particular type of crop. It includes consumptive use, evaporation, and seepage from ditches and canals, and the water eventually returned to streams by percolation and surface runoff, usually expressed in acre-feet per acre.

**Effective Precipitation:** The amount of rain that falls during the growing season and is available for growth of crops. Effective precipitation is a portion of the total rain that falls during the growing season and is a function of the type of soil, the time period in which each rain falls, and its intensity. Effective precipitation is usually less than precipitation measured at a given point.

**Enlargement:** A subsequent right awarded to a ditch or structure enlarging the amount granted originally. More than one enlargement may be awarded to a ditch or structure and each enlargement will have a priority related to the date it was appropriated and applied to beneficial use. Enlargements may be absolute or conditional.

**Entire Water:** A term used to describe water which occupies volume to the exclusion of everything else. Water flowing in a canal or a river is entire water. Interstitial water occupies only the interstices between grains in an aquifer.

**Evaporation:** The physical process by which a liquid or solid is transformed to the gaseous state which, in irrigation, usually is restricted to the change of water from liquid to gas.

**Evapotranspiration:** The combined processes by which water is transferred from the earth surface to the atmosphere; evaporation of liquid or solid water plus transpiration from plants (See Consumptive Use).

**Floodplain:** An area adjacent to a stream or other water course which is subject to flooding.

**Flowing Well:** A well from an artesian aquifer in which the water is under sufficient pressure to rise above the ground surface.

**Forteiture:** The loss of a water right based on its nonuse for a statutorily provided period of time.

**Futile Call:** A situation in which a junior priority will be permitted to continue to divert in spite of demands by a senior appropriator in the same water shed, because to curtail the junior from diversion would not effectively produce water for beneficial use for the senior.

**Gage:** (1) An instrument used to measure magnitude or position; gages may be used to measure the elevation of a water surface, the velocity of flowing water, the pressure of water, the amount of intensity of precipitation, the depth of snowfall, and so on.(2) The act or operation of registering or measuring magnitude or position. (3) The operation, including both field and office work, of measuring the discharge of a stream of water in a waterway.

**Gage Height:** The height of the water surface above the gage datum. Gage height is often used interchangeably with the more general term, stage, although gage height is more appropriate when used with a gage reading.

**Gaging Station:** A particular site on a stream, canal, lake, or reservoir where systematic observations of gage height or discharge are made.

**Geohydrology:** The branch of hydrology relating to subsurface or subterranean waters.

**Gradient:** A slope of the water table tending to cause the flow of groundwater.

**Groundwater:** Groundwater is usually defined as any water not visible on the surface of the ground under natural conditions.

**Groundwater Basin:** A physiographic or geologic unit containing at least one aquifer of significant areal extent.

**Groundwater, Confined:** Groundwater under pressure significantly greater than atmospheric, with its upper limit the bottom of a bed with hydraulic conductivity distinctly lower than that of the material in which the confined water occurs.

**Groundwater Divide:** A line on a water table on either side of which the water table slopes downward. It is analogous to a drainage divide between two drainage basins on a land surface.

**Groundwater, Free:** Unconfined groundwater whose upper boundary is a free water table.

**Groundwater Hydrology:** The branch of hydrology that treats groundwater, its occurrence and movements, its replenishment and depletion, the properties of rocks that control groundwater movement and storage, and the methods of investigation and utilization of groundwater.

**Groundwater Mining:** The pumping of groundwater from a basin where the safe yield is very small, thereby extracting groundwater which accumulated over a long period of time. It occurs when withdrawals exceed replenishment or when replenishment is negligible.

**Groundwater Outflow:** The part of the discharge from a drainage basin that occurs through the groundwater. The term underflow is often used to describe the groundwater outflow that takes place in valley alluvium (instead of the surface channel) and thus is not measured at a gaging station.

**Groundwater Overdraft:** Pumpage of groundwater in excess of safe yield.

**Groundwater, Perched:** Groundwater that is separated from the main body of groundwater by unsaturated material.

**Groundwater Recharge:** Inflow to a groundwater reservoir.

**Groundwater Reservoir:** An aquifer or aquifer system in which groundwater is stored. The water may be placed in the aquifer by either artificial or natural means.

**Growing Season:** That portion of the year, usually May through October, that the plants are consuming water and nutrients.

**Headgate:** A physical structure on a stream through which water is diverted into a ditch.

**Historic Use:** The documented diversion and use of water by a water right holder in a ditch over a period of years.

**Image:** A hypothetical well, source, or sink used as a mathematical device to satisfy a boundary condition.

**Impermeable:** Not permeable.

**Impervious:** An adjective describing a material through which water either cannot pass or through which it passes with great difficulty.

**Inch:** A Colorado statutory inch is defined as follows:

Every inch shall be considered equal to an inch square orifice under a five-inch pressure, and a five-inch pressure shall be from the top of the orifice of the box put into the banks of the ditch, to the surface of the water; said boxes, or any slot or apperture through which such water may be measured, shall in all cases be six inches perpendicular, inside measurement, except boxes delivering less than twelve inches, which may be square and said box put into the banks of ditch shall have a descending grade from the water in the ditch or not less than one-eighth of an inch to the foot. See also Miners Inch.

**Infiltration:** Water moving into the ground from a surface supply such as precipitation or irrigation. Infiltration rates are reckoned on the basis that the water is entire water.

**Initial Conditions:** The conditions that prevail at the time of initiation of a transient.

**Insolation:** (Contracted from incoming solar radiation.) Solar radiation received at the earth's surface.

**Instream Flow Needs:** Those habitat requirements within the running water ecosystem related to current velocity and depth which present the optimum conditions of density (or diversity) or physiological stability to the aquatic organism being examined.

**Instream Use:** Any use of water which does not require diversion from a water course or impoundment.

**Intent to Appropriate:** That condition in which the mind has formed a definite plan to divert or store water, create the means to do so, and apply such water to a beneficial use.

**Interstate:** Having to do with two or more states.

**Interstitial Water:** Water which occupies the interstices between grains in a permeable bed. (See Entire Water.)

**Intrastate:** Having to do with matters confined to one state.

**Irrigated Area:** The area upon which water is artificially applied for the production of crops.

**Irrigation:** The application of water to crops, lawns, and gardens by artificial means to supplement natural precipitation. Water can be applied by spreading, sprinkling, or dripping.

**Irrigation Efficiency:** The ratio of the volume of water consumed by a specific beneficial use as compared to the volume of water delivered. Efficiency may be computed in terms of the water diverted at the ditch headgate or the water delivered to the farm headgate.

**Irrigation Return Flow:** Applied water which is not consumptively used and returns to a surface water or groundwater supply. In water right litigation the definition may be restricted to measureable water returning to the stream from which it was derived.

**Irrigation, Supplemental:** An additional irrigation water supply which supplements the initial or primary supply.

**Irrigation Water Requirement:** The quantity of water, exclusive of effective precipitation, that is required for various beneficial uses.

**Isohyet:** A line on the surface of the earth, as represented on a map, connecting all points of equal precipitation. Also called isohyetal line and isopluvial line.

**Lateral:** A minor ditch headgating off the main ditch used to direct water onto the land. A ditch may have many laterals, depending on the amount of acreage irrigated, the slope of the land, and the rate of seepage losses.

**Law of the River:** The name applied to the legal framework comprised of interstate and interregional compacts, state and federal laws, Supreme Court decisions, and international treaties which govern the distribution of water from the Colorado River system.

**Linearization:** Many of the differential equations representing physical relationships are inherently nonlinear in form. Such relationships are generally difficult to handle and it is often desirable to replace them with approximations which have a more tractable linear form. This may be done in specific cases, by neglect of small quantities, by replacing a curve with its tangent in a range of interest or by other means. The process is called linearization.

**Line Source:** A source uniformly distributed along a line.

**Local Convergence:** A convergence of flow of groundwater as to a drain tile.

**Loss:** The difference between the amount of water that is actually placed on the land and the amount of water that was physically diverted to the headgate. Losses usually are from seepage and evaporation.

**Lysimeter:** An instrument used to measure the quantity or rate of downward water movement through a block of soil, usually undisturbed, or to collect such percolated water for analysis of its quality.

**Meinzer Unit:** A unit of permeability used in the older publications of the U.S. Geological Survey. It is defined as the flow of water in gallons per day through a cross-sectional area of 1 square foot under a gradient of 1 foot head of water per foot of length, measured in the direction of flow, at a temperature of 60°F.

**Miner's Inch:** The term miner's inch, formerly used in hydraulic mining and irrigation in the western United States, is practically obsolete. It is defined as the quantity of water which will flow through an orifice 1 inch square under a stated head which varies from 4 to 6½ inches in different localities. The use of this unit has led to much confusion; its value in terms of cubic feet per second has been fixed by statute in most of the western states, as follows:

(1) 50 miner's inches = 1 second foot in Idaho, Kansas, Nebraska, New Mexico, North Dakota, South Dakota, Utah, Washington, and Southern California.

(2) 40 miner's inches = 1 second foot in Arizona, Montana, Nevada, Oregon, and Northern California.

(3) 38.4 miner's inches = 1 second foot in Colorado.

**Nonconsumptive Use:** A use of water that does not reduce the supply, such as for hunting, fishing, boating, water-skiing, and swimming.

**Nontributary Groundwater:** Water that is not part of a natural stream as established through geologic and hydrologic facts. The factual determination of nontributary usually involves the length of time the impact of withdrawal would take to reach the stream and the amount of impact relative to the total volume of surface flow impacted.

**Original Right:** The first right awarded to a ditch or storage structure.

**Paper Right:** A document purporting to be legal proof of a water right, but which has lost its legal validity because of abandonment or lack of due diligence in perfecting the right.

**Parallel Drains:** Drains of the type installed for drainage of agricultural land. They can be open ditches or buried tile lines.

**Perfection of a Water Right:** The process of meeting all of the legal requirements for establishing a legal right to the use of water. Once perfected, a conditional water right becomes absolute.

**Permeability:** A term used to describe the ability of water or other liquid to move through a porous formation under the action of a gradient. The facility with which a fluid will move through a formation is greater for some than for others. For a given bed, the permeability is expressed by a constant K representing the flow through unit are in unit time under the influence of a unit gradient. The flow is expressed in terms of entire water.

**Phreatophyte:** A water-loving plant which consumes a substantial amount of water.

**Potential Evapotranspiration:** The rate at which water, if available, would be removed from the soil and plant surface expressed as the rate of latent heat transfer per square centimeter or depth of water. For comparative purposes potential evapotranspiration refers to a well-watered crop like alfalfa (lucerne) with 30 to 50 centimeters of top growth and about 100 millimeters of fetch under given climatic conditions unless otherwise defined.

**Prior Appropriation:** A term describing the general process by which water rights are distributed among several claimants. In the West the first person to use the water beneficially gets the water right, whether or not that person owns land next to the river or lake from which the water is diverted.

**Priority:** The relative seniority of a water right as determined by its adjudication date and appropriation date. Other factors are sometimes involved in determining priority. The priority of a water right determines its ability to divert in relation to other rights in periods of limited supply.

**Production (Well):** The total volume of well flow counted from the time of initiation of flow.

**Radial Flow:** Flow converging toward a center.

**Referee:** A person selected by the water judge to carry out certain judicial functions of the water court.

**Reservoir:** A pond, lake, or basin, either natural or artificial, used for the storage, regulation, and control of water.

**Return Flow:** Unconsumed water which returns to its source or some other water body after its diversion as surface water or its extraction from the ground. Also, tailwater, drainage.

**Riparian:** Pertaining to the banks of a stream, lake, or body of water.

**Riparian Land:** Land which abuts upon the banks of a stream or other natural body of water.

**Riparian Rights:** A system used primarily in the eastern states to determine who has rights to water. The riparian system gives water rights to the owners of the lands through which water flows. (See Prior Appropriation)

**Riparian Vegetation:** Vegetation growing on the banks of a stream or other body of surface water.

**River Basin:** The area drained by a river and its tributaries.

**River Stage:** The elevation of the water surface at a specified station above some arbitrary zero datum.

**Root:** In mathematic expressions, a value of an argument which will cause some given function of the argument to pass through zero.

**Runoff:** Precipitation that flows to and in surface streams; renewable water.

**Salinity:** The amount of dissolved solids in water, sometimes referred to as Total Dissolved Solids (TDS), as well as Soluble Mineral Content (SMC). 500 ppm is acceptable for drinking water; plant damage occurs at 800–1000 ppm.

**Salvaged Water:** Water which is saved to a natural stream by human modification of natural conditions.

**Seepage:** (1) The slow movement of water through small cracks, pores, interstices of a material into or out of a body of surface or subsurface water. (2) The loss of water by infiltration into the soil from a canal, reservoir, or other body of water, or from a field. Seepage is generally expressed as flow volume per unit time. During the process of priming, the loss is called absorption loss.

**Slurry:** A mixture of liquid and solid, typically, water and coal mixed together for transportation purposes.

**Solar Radiation:** The total electromagnetic radiation emitted by the sun (see Insolation).

**Specific Yield:** The amount of water that rock or soil, after being saturated, will yield when drained by gravity, expressed as a ratio or as a percentage of the volume of the rock or soil.

**Staff Gage:** A graduated scale used to indicate the height of the water surface in a stream channel, reservoir, lake, or other water body.

**Stage:** The height of a water surface above an established datum plane (see Gage Height).

**State Engineer:** The chief executive officer in the executive department of the state government who administers water rights.

**Steady State Cases:** These are groundwater or surface water conditions which do not change with time.

**Storage Right:** A right defined in terms of the volume of the water which may be diverted from the flow of the stream and stored in a reservoir or lake to be released and used at a later time either within the same year or a subsequent year.

**Stream Depletion (Wells):** A depletion of a streamflow caused by the operation of wells installed in an aquifer hydraulically connected to the stream.

**Substantive Law:** The law defining rights to the use of water.

**Total Consumptive Use:** The amount of water, regardless of its source, used by the crops during the growing season. It is the amount of water that is physically removed from the stream's system and is not available for other users on the stream.

**Trans-Basin Diversion:** The removal of the water of a natural stream from its natural basin into the natural basin of another stream.

**Transfer:** The process of moving a water right originally decreed to one ditch, to another ditch or point of diversion, by court decree. A transferred water right generally retains its priority in the stream system and may or may not retain its right to divert its entire decreed amount.

**Transient Cases:** Conditions which are changing with time.

**Transpiration:** The process by which water in plants is transferred as water vapor to the atmosphere.

**Tributary Drainage:** The area from which water drains by gravity into a water course.

**Tributary Groundwater:** Seepage, underflow, and percolating water that will eventually become part of the natural stream. A natural stream's waters include water in the unconsolidated alluvial aquifer of sand, gravel, and other sedimentary materials, and all other waters hydraulically connected thereto, which can influence the rate or direction of movement of the water in that alluvial aquifer or natural stream. In Colorado, all groundwater is presumed to be tributary unless proved otherwise.

**Turbulence:** A state of fluid flow in which instantaneous velocities exhibit irregular and apparently random fluctuations.

**Unconfined Aquifer:** An aquifer in which the water table serves as the upper surface of the zone of saturation.

**Unit Consumptive Use (Irrigation):** The amount of water used by crops for growth, less effective precipitation, expressed in acre-feet per acre or feet of water. Unit consumptive use is considered synonymous with irrigation consumptive use and is less than total consumptive use. Water for consumptive use may be supplied from surface water diverted by a ditch and groundwater occurring naturally beneath the crops.

**Vapor Pressure:** The partial pressure of water vapor in the atmosphere.

**Virgin Flow:** The flow of a river that would occur in the absence of human activities; synonymous with native supply.

**Void Ratio:** The ratio of the volume of drainable or fillable voids to the gross volume.

**Voir Dire:** The in-court examination of an expert, to determine whether or not he is qualified to express an opinion on a particular subject, or an exhibit to determine whether or not it is admissible.

**Volume:** A specific quantity of water generally expressed in terms of acre-feet. An acre-foot is defined as the amount of water required to cover 1 acre of land to a depth of 1 foot and is equivalent to 43,560 cubic feet, or 325,850 gallons.

**Water Commissioner:** Public officials under the direction of the division engineers who carry out the detailed daily administration of the waters of portions of each water division.

**Water Course:** A place on the earth's surface where water flows, regularly or intermittently, in a defined channel.

**Water Court:** In Colorado, special division of a district court with a district judge, called the water judge, to deal with certain specific water matters principally having to do with adjudication and change of water rights.

**Water Development:** The process of building diversion, storage, pumping, and/or conveyance facilities to apply water to beneficial use.

**Water District:** A subdivision of a water division.

**Water Division:** A major watershed of the state.

**Water Right:** A right to use, in accordance with its priority, a certain portion of the waters of the state by reason of the appropriation of the same.

**Watershed:** The area from which water drains to a single point.

**Water Table:** The upper limit of the completely saturated material in an aquifer.

**Water Well:** A water well is a hole or shaft, usually vertical, excavated in the earth for bringing groundwater to the surface. Occasionally wells serve other purposes, such as subsurface exploration and observation, artificial recharge, and disposal of wastewaters. Many methods exist for constructing wells; selection of a particular method depends on the purpose of the well, the quantity of water required, depth to groundwater, geologic conditions, and economic factors. Shallow wells are dug, bored, driven or jetted; deep wells are drilled by cable tool or rotary methods. Colorado Statutes define a well as any excavation that is drilled, cored, bored, washed, driven, dug, jetted, or otherwise constructed, when the intended use of such excavation is for the location, diversion, artificial recharge, or acquisition of groundwater, but such term does not include an excavation made for the purpose of obtaining or for prospecting for oil, natural gas, minerals, or products of mining or quarrying, or for inserting media to repressure oil or natural gas bearing formations or for storing petroleum, natural gas, or other products.

**Water Year:** The 12-month period October 1st through September 30th. The water year is designated by the calendar year in which it ends and which includes 9 of the 12 months. Thus, the year ending September 30, 1959, is the 1959 water year.

**Well:** See Water Well.

**Yield:** (1) The quantity of water expressed either as a continuous rate of flow or as a volume per unit of time (af per year), which can be collected for a given use or uses from surface or groundwater sources on a watershed. The yield may vary with the proposed use, the plan of development, and also economic considerations. Yield is fairly synonymous with water crop. (2) Total runoff. (3) The streamflow in a given interval of time derived from a unit area of watershed. It is determined by dividing the observed streamflow at a given location by the drainage area above that location and is usually expressed in cubic feet per second per square mile. See also Yield, Firm; Yield Perennial; Yield, Safe.

**Yield, Average Annual:** The average annual supply of water produced by a given stream or water development.

**Yield, Firm:** The maximum annual supply of a given water development that is expected to be available on demand, with the understanding that lower yields will occur in accordance with a predetermined schedule or probability.

**Yield, Perennial:** The amount of usable water of a groundwater reservoir that can be economically withdrawn and consumed each year for an indefinite period of time. It cannot exceed the natural recharge to that groundwater reservoir and ultimately is limited to the maximum amount of discharge that can be utilized for beneficial use.

**Yield, Safe:** With reference to either a surface or groundwater supply, the rate of diversion or extraction for consumptive use which can be maintained indefinitely, within the limits of economic feasibility, under specified conditions of water supply development (see also Yield, Perennial).

**Yield, Water Right:** The volume of water diverted by a water right. Yield may be expressed as an average for a period of years (average yield) or as the yield of one selected year representing the lowest or critical amount of water provided (critical year yield). Yield also may refer to diversion at the headgate (headgate yield) or at the farm turnout where it is applied to irrigation (farm yield). The difference between headgate yield and farm yield is the amount of water lost to seepage and other causes related to the conveyance of water through the ditch.

**Zanjero:** Spanish term for ditch rider or water commissioner. One who regulates the delivery of water to irrigation users.

# APPENDIX

# CONVERSION FACTORS

## CONVERSION FACTORS BY PHYSICAL QUANTITY

| Multiply | By | To Obtain |
|---|---|---|
| **Length** | | |
| Miles | 1.60935 | Kilometers |
| Miles | 1,760 | Yards |
| Miles | 5,280 | Feet |
| Miles | 63,360 | Inches |
| Meters | 0.00062137 | Miles |
| Meters | 1.0936 | Yards |
| Meters | 3.28088 | Feet |
| Meters | 39.37 | Inches |
| Meters | 100 | Centimeters |
| Meters | 0.001 | Kilometers |
| Yards | 0.9144 | Meters |
| Yards | 0.00056818 | Miles |

(*continued*)

| Multiply | By | To Obtain |
|---|---|---|
| Yards | 3.0 | Feet |
| Yards | 36 | Inches |
| Feet | 0.3048 | Meters |
| Feet | 0.00018939 | Miles |
| Feet | 0.33333 | Yards |
| Feet | 12 | Inches |
| Inch | 0.08333 | Feet |
| Inch | 0.027778 | Yards |
| Inch | 0.000015783 | Miles |
| Inch | 2.54 | Centimeters |

## Surface

| Multiply | By | To Obtain |
|---|---|---|
| Square Mile | 4,014,489,600 | Square inches |
| Square Mile | 27,878,400 | Square feet |
| Square Mile | 3,097,600 | Square yards |
| Square Mile | 640 | Acres |
| Square Mile | 259 | Hectares |
| Acre | 208.71 | Feet square |
| Acre | 0.404687 | Hectares |
| Acre | 0.0015625 | Square miles |
| Acre | 4,840 | Square yards |
| Acre | 43,560 | Square feet |
| Acre | 4,047 | Square meters |
| Square Yards | 0.83613 | Square meters |
| Square Yards | 0.0000003228 | Square miles |
| Square Yards | 0.0002066 | Acres |
| Square Yards | 9 | Square feet |
| Square Yards | 1,296 | Square inches |
| Square Feet | 0.092903 | Square meters |
| Square Feet | 0.000000003587 | Square miles |
| Square Feet | 0.000022957 | Acres |
| Square Feet | 0.11111 | Square yards |
| Square Feet | 144 | Square inches |
| Square inches | 0.0000000002491 | Square miles |
| Square inches | 6.45163 | Square centimeters |
| Square inches | 0.0000001594 | Acres |
| Square inches | 0.0007716 | Square yards |
| Square inches | 0.006944 | Square feet |

## Volume

| Multiply | By | To Obtain |
|---|---|---|
| Acre-feet | 325,851 | U.S. gallons |
| Acre-feet | 43,560 | Cubic feet |

| Multiply | By | To Obtain |
|---|---|---|
| Acre-feet | 1,613.3 | Cubic yards |
| Acre-feet | 1,233.49 | Cubic meters |
| Cubic yard | 27 | Cubic feet |
| Cubic yard | 46,656 | Cubic inches |
| Cubic yard | 0.00061983 | Acre-feet |
| Cubic yard | 0.76456 | Cubic meters |
| Cubic feet | 1,728 | Cubic inches |
| Cubic feet | 7.4805 | U.S. gallons |
| Cubic feet | 28.317 | Liters |
| Cubic feet | 0.037037 | Cubic yards |
| Cubic feet | 0.000022957 | Acre-feet |
| U.S. gallon | 231 | Cubic inches |
| U.S. gallon | 3.78543 | Liters |
| U.S. gallon | 0.13368 | Cubic feet |
| U.S. gallon | 0.00000307 | Acre-feet |
| Cubic inches | 16.3872 | Cubic centimeters |
| Cubic inches | 0.004329 | U.S. gallons |
| Cubic inches | 0.0005787 | Cubic feet |
| Second feet | 448.8 | U.S. gallons/minute |
| Second feet | 60 | Cubic feet/minute |
| Second feet | 3,600 | Cubic feet/hour |
| Second feet | 86,400 | Cubic feet/day |
| Second feet | 723.9669 | Acre-feet per year |
| Second feet | 1.9835 | Acre-feet per day |
| Second feet | 0.9917 | Acre-inch per hour |
| Second feet | 50 | Miner's inch in Idaho, Kansas, Nebraska, South Dakota, North Dakota, New Mexico, Utah, Washington and Southern California |
| Second feet | 40 | Miner's inch in Arizona, Montana, Oregon, Nevada, and Northern California |
| Second feet | 38.4 | Miner's inch in Colorado |
| Second feet | 0.028317 | Cubic meters per second |
| Second feet | 1.699 | Cubic meters per minute |
| Second feet | 101.941 | Cubic meters per hour |
| Second feet | 2446.58 | Cubic meters per day |
| Million gallons per day | 1.547 | Second-feet |
| Million gallons per day | 3.07 | Acre-feet per day |
| Million gallons per day | 2.629 | Cubic meters per minute |
| Feet per second | 0.68 | Miles per hour |
| Feet per second | 1.097 | Kilometers per hour |
| Feet per second | 30.48 | Centimeters per second |

(*continued*)

| Multiply | By | To Obtain |
|---|---|---|
| Pounds, avoirdupois | 0.4536 | Kilograms |
| Pounds, avoirdupois | 16 | Ounces |
| Pounds, avoirdupois | 7,000 | Grains |
| Pounds, avoirdupois | 1.21528 | Pounds, troy |
| Kilograms | 1,000 | Grams |
| Kilograms | 15.432 | Grains |
| Kilograms | 2.2046 | Pounds, avoirdupois |
| Atmosphere, at sea level | 76.0 | Centimeter of mercury |
| Atmosphere, at sea level | 29.92 | Inches of mercury |
| Atmosphere, at sea level | 33.90 | Feet of water |
| Atmosphere, at sea level | 14.70 | Pounds per square inch |
| Degrees | 60 | Minutes |
| Degrees | 3,600 | Seconds |
| Degrees | 0.01745 | Radians |
| Horsepower | 33,000 | Foot-pounds per minute |
| Horsepower | 550 | Foot-pounds per second |
| Horsepower | 0.7457 | Kilowatts |
| B.t.u. | 778 | Foot-pounds |

1 second-foot falling 8.81 feet = 1 horsepower.
1 second-foot falling 10.0 feet = 1.135 horsepower.
1 second-foot falling 11.0 feet = 1 horsepower at 80% efficiency.
1 second-foot flowing for 1 year will cover 1 square mile 1.131 feet, or 13.572 inches, deep.
1 inch depth of water on 1 square mile = 2,323,200 cubic feet = 0.0737 second-foot for 1 year.

## CONVERSION FACTORS FROM ENGLISH TO METRIC UNITS

| Multiply | By | To Obtain |
|---|---|---|
| **Length** | | |
| Inches | 25.4 (exactly) | Millimeters |
| Feet | 30.48 (exactly) | Centimeters |
| Miles | 1.609344 (exactly) | Kilometers |
| **Area** | | |
| Square inches | 6.4516 (exactly) | Square centimeters |
| Square feet | 0.092903 (exactly) | Square meters |
| Acres | 0.0040469 | Square kilometers |
| Square miles | 2.58999 | Square kilometers |

| Multiply | By | To Obtain |
|---|---|---|
| **Volume** | | |
| Cubic feet | 0.0283168 | Cubic meters |
| | 28.3168 | Liters |
| Gallons (U.S.) | 3.78543 | Liters |
| Cubic yards | 0.7646 | Cubic meters |
| Acre-feet | 1233.5 | Cubic meters |
| **Mass** | | |
| Pounds, avoirdupois | 0.45359237 (exactly) | Kilograms |
| Tons (2,000 pounds) | 907.185 | Kilograms |
| **Acceleration** | | |
| Feet per second per second | 0.3048 | Meters per second per second |
| **Force/Unit Area** | | |
| Pounds per square inch | 0.070307 | Kilograms per square centimeter |
| Pounds per square foot | 4.88243 | Kilograms per square meter |
| Feet of water column (at 20°C) | 2.246 | Centimeters of mercury column |
| | 0.03041 | Kilograms per square centimeter |
| **Mass/Volume (Density)** | | |
| Pounds per cubic foot | 16.0185 | Kilograms per cubic meters |
| | 0.0160185 | Grams per cubic centimeter |
| **Velocity** | | |
| Feet per second | 30.48 (exactly) | Centimeters per second |
| Inches per hour | 2.540 (exactly) | Centimeters per hour |
| Feet per year | 0.3048 (exactly) | Meters per year |
| **Flow** | | |
| Cubic feet per second | 0.028317 | Cubic meters per second |
| | 28.317 | Liters per second |
| Cubic feet per minute | 0.4719 | Liters per second |
| Gallons per minute | 0.06309 | Liters per second |
| | 3.7854 | Liters per minute |

(*continued*)

| Multiply | By | To Obtain |
|---|---|---|
| | **Power** | |
| Horsepower (British) (defined 550 ft lb/sec) | 745.700 | Watts |
| | 1.014 | Horsepower (metric) (defined 75 kg-m/sec) |
| | **Seepage** | |
| Cubic feet per square foot per day | 304.8 | Liters per square meter per day |
| | **Viscosity** | |
| Dynamic viscosity (pound second per square foot) | 4.8824 | Kilogram second per square meter |
| Kinematic viscosity (square feet per second) | 0.092903 (exactly) | Square meters per second |
| | **Surface Tension** | |
| Pounds per foot | 1.4882 | Kilograms per meter |
| | **Gas Constant** | |
| Feet per degree F | 0.5486 | Meters per degree Celsius* |

Source: U.S. Bureau of Reclamation

* For all practical purposes, the Celsius and centigrade scales are synonymous.

# BIBLIOGRAPHY

American Society of Civil Engineers, Technical Committee on Irrigation Water Requirements, Irrigation and Drainage Division. *Consumptive Use of Water and Irrigation Water Requirements.* New York: The Society, 1973.

Bird, John W. *Origin and Growth of Federal Reserved Water Rights.* ASCE Journal of the Irrigation and Drainage Division. March 1981.

Burges, Stephen J., and Reza, Maknoon. *A Systematic Examination of Issues in Conjunctive Use of Ground and Surface Waters.* Technical Report No. 44, Department of Civil Engineering. Seattle: University of Washington. September 1975.

*Colorado Water Law Practice.* Program of Advanced Professional Development. Denver, Colorado: University of Denver. 1976.

*Critical Water Problems Facing the Eleven Western States, Westwide Study.* Three Volumes. Department of the Interior, U.S. Government Printing Office, Washington D.C. April 1975.

*The Denver Basin: Should it be Designated?* Continuing Legal Education, Inc. Denver, Colorado: Continuing Legal Education, 1982.

*Denver Law Journal.* Volume 47, Number 2. Denver, Colorado: University of Denver, 1970.

Dewsnut, Richard L. and Dallin W. Jensen, editors. *A Summary Digest of State Water Laws.* National Water Commission. Washington, D.C.: U.S. Government Printing Office, 1973.

"The Divining Rod: A History of Water Witching." *Water Supply Paper 416*. Department of the Interior. Washington, D.C.: United States Geological Survey, 1917.

Environmental Resource Center. *Colorado Water Laws, A Compilation of Statutes, Regulations, Compacts and Selected Cases.* Comp. by George Radosevich. 3 Vol. Information Series Number 17. Fort Collins: Colorado State University, 1975.

Getches, David H. *Water Law in a Nutshell.* St. Paul: West Publishing Co., 1984.

Glover, Robert E. *The Pumped Well.* Technical Bulletin 100. Fort Collins: Colorado State University, 1968.

Glover, Robert E. *Transient Groundwater Hydraulics.* 2nd Printing. Fort Collins, Colorado: Water Resources Publications, 1978.

*Groundwater and Wells.* UOP Inc., Johnson Division. St. Paul: UOP Inc., 1966.

Harris, Linda G. *New Mexico Water Rights.* Miscellaneous Report No. 15. Albuequerque: New Mexico Water Resources Research Institute, March 1984.

Harrison, David L. and Jeris A. Danielson. *Using An Expert in a Groundwater Case: The San Luis Valley Case.* Short Course on Groundwater Allocation, Development, and Pollution. Natural Resources Law Center. Boulder: University of Colorado School of Law, June 1983.

Houghy, James E. "The Engineer as Expert Witness." *Civil Engineering.* New York: American Society of Civil Engineers, December 1981.

Hutchins, Wells A. and others. "Water Rights Laws in the Nineteen Western States." *Miscellaneous Publication No. 1206.* Natural Resource Economics Division, U.S. Department of Agriculture. Washington, D.C.: U.S. Government Printing Office, 1977.

"Irrigation Water Requirements." *Technical Release No. 21.* Soil Conservation Service, U.S. Department of Agriculture. - Washington, D.C.: U.S. Government Printing Office, Rev. 1970.

Law, James P. *National Irrigation Return Flow Research and Development Program.* Ada: Environmental Protection Agency, December 1971.

Myers, Charles J. and A. Dan Tarlock. *Water Resource Management: A Coursebook in Law and Public Policy.* Mineola, New York: The Foundation Press, 1971.

National Academy of Sciences. National Research Council, Committee on Water. *Water and Choice in the Colorado Basin, An Example of Alternatives in Water Management.* Washington, D.C.: National Academy of Sciences, 1968.

*The Nation's Water Resources.* United States Water Resources Council. Washington, D.C.: U.S. Government Printing Office, 1968.

Radosevich, George and others. *Evolution and Administration of Colorado Water Law.* Fort Collins, Colorado: Water Resource Publications, 1976.

*Report of Groundwater Legislation Committee.* Denver, Colorado: Department of Natural Resources, August 1984.

Stefferud, Alfred, Ed. *Water: The Yearbook of Agriculture, 1955.* United States Department of Agriculture. Washington, D.C., 1955.

Trelease, Frank J. *Cases and Materials on Water Law.* St. Paul: West Publishing Co., 1984.

Viessman, Warren and others. *Introduction to Hydrology.* New York: Thomas J. Crowell Company, 1972.

*Water in the West.* American Institute of Professional Geologists. Evergreen, Colorado: AIPG, November 1981.

*Water Law for the NonSpecialist Practitioner.* Continuing Legal Education in Colorado. Denver, Colorado: University of Denver, 1974.

*Water Policies for the Future*. Final Report to the President and to the Congress of the United States by the National Water Commission. Washington, D.C.: U.S. Government Printing Office, June, 1973.

*Wyoming Water and Irrigation Laws*. Comp. by George L. Christopulos, State Engineer. Cheyenne, Wyoming, 1975.

# INDEX